SpringerBriefs in Computer Science

Series Editors

Stan Zdonik, Brown University, Providence, RI, USA

Shashi Shekhar, University of Minnesota, Minneapolis, MN, USA

Xindong Wu, University of Vermont, Burlington, VT, USA

Lakhmi C. Jain, University of South Australia, Adelaide, SA, Australia

David Padua, University of Illinois Urbana-Champaign, Urbana, IL, USA

Xuemin Sherman Shen, University of Waterloo, Waterloo, ON, Canada

Borko Furht, Florida Atlantic University, Boca Raton, FL, USA

V. S. Subrahmanian, University of Maryland, College Park, MD, USA

Martial Hebert, Carnegie Mellon University, Pittsburgh, PA, USA

Katsushi Ikeuchi, University of Tokyo, Tokyo, Japan

Bruno Siciliano, Università di Napoli Federico II, Napoli, Italy

Sushil Jajodia, George Mason University, Fairfax, VA, USA

Newton Lee, Institute for Education, Research and Scholarships, Los Angeles, CA, USA

SpringerBriefs present concise summaries of cutting-edge research and practical applications across a wide spectrum of fields. Featuring compact volumes of 50 to 125 pages, the series covers a range of content from professional to academic.

Typical topics might include:

- A timely report of state-of-the art analytical techniques
- A bridge between new research results, as published in journal articles, and a contextual literature review
- A snapshot of a hot or emerging topic
- An in-depth case study or clinical example
- A presentation of core concepts that students must understand in order to make independent contributions

Briefs allow authors to present their ideas and readers to absorb them with minimal time investment. Briefs will be published as part of Springer's eBook collection, with millions of users worldwide. In addition, Briefs will be available for individual print and electronic purchase. Briefs are characterized by fast, global electronic dissemination, standard publishing contracts, easy-to-use manuscript preparation and formatting guidelines, and expedited production schedules. We aim for publication 8–12 weeks after acceptance. Both solicited and unsolicited manuscripts are considered for publication in this series.

**Indexing: This series is indexed in Scopus, Ei-Compendex, and zbMATH **

Ting Wang • Bo Li • Mingsong Chen • Shui Yu

Machine Learning Empowered Intelligent Data Center Networking

Evolution, Challenges and Opportunities

 Springer

Ting Wang 🆔
Software Engineering Institute
East China Normal University
Shanghai, China

Mingsong Chen
Software Engineering Institute
East China Normal University
Shanghai, China

Bo Li 🆔
Software Engineering Institute
East China Normal University
Shanghai, China

Shui Yu 🆔
School of Computer Science
University of Technology Sydney
Ultimo, NSW, Australia

ISSN 2191-5768 ISSN 2191-5776 (electronic)
SpringerBriefs in Computer Science
ISBN 978-981-19-7394-9 ISBN 978-981-19-7395-6 (eBook)
https://doi.org/10.1007/978-981-19-7395-6

This Springer imprint is published by the registered company Springer Nature Singapore Pte Ltd.
The registered company address is: 152 Beach Road, #21-01/04 Gateway East, Singapore 189721,
Singapore

Preface

To support the needs of ever-growing cloud-based services, the number of servers and network devices in data centers is increasing exponentially, which in turn results in high complexities and difficulties in network optimization. To address these challenges, both academia and industry turn to artificial intelligence technology to realize network intelligence. To this end, a considerable number of novel and creative machine learning-based (ML-based) research works have been put forward in recent few years. Nevertheless, there are still enormous challenges faced by the intelligent optimization of data center networks (DCNs), especially in the scenario of online real-time dynamic processing of massive heterogeneous services and traffic data. To the best of our knowledge, there is a lack of systematic and original comprehensive investigations with in-depth analysis on intelligent DCN. To this end, in this book, we comprehensively investigate the application of machine learning to data center networking and provide a general overview and in-depth analysis of the recent works, covering flow prediction, flow classification, load balancing, resource management, energy management, routing optimization, congestion control, fault management, and network security. In order to provide a multi-dimensional and multi-perspective comparison of various solutions, we design a quality assessment criteria called REBEL-3S to impartially measure the strengths and weaknesses of these research works. Moreover, we also present unique insights into the technology evolution of the fusion of data center networks and machine learning, together with some challenges and potential future research opportunities.

Shanghai, P.R. China
March 2022

Ting Wang
Bo Li
Mingsong Chen
Shui Yu

Acknowledgments

This work was partially supported by the grants from National Key Research and Development Program of China (2018YFB2101300), Natural Science Foundation of China (61872147), the Dean's Fund of Engineering Research Center of Software/Hardware Co-design Technology and Application, Ministry of Education (East China Normal University), and the grants from Shenzhen Science and Technology Plan Project (CJGJZD20210408092400001).

Contents

1	**Introduction**	1
	References	5
2	**Fundamentals of Machine Learning in Data Center Networks**	9
	2.1 Learning Paradigm	9
	2.2 Data Collection and Processing	10
	2.2.1 Data Collection Scenarios	11
	2.2.2 Data Collection Techniques	11
	2.2.3 Feature Engineering	11
	2.2.4 Challenges and Insights	12
	2.3 Performance Evaluation of ML-Based Solutions in DCN	13
	References	14
3	**Machine Learning Empowered Intelligent Data Center Networking**	15
	3.1 Flow Prediction	15
	3.1.1 Temporal-Dependent Modeling	16
	3.1.2 Spatial-Dependent Modeling	16
	3.1.3 Discussion and Insights	17
	3.2 Flow Classification	17
	3.2.1 Supervised Learning-Based Flow Classification	22
	3.2.2 Unsupervised Learning-Based Flow Classification	23
	3.2.3 Deep Learning-Based Flow Classification	23
	3.2.4 Reinforcement Learning-Based Flow Classification	24
	3.2.5 Discussion and Insights	24
	3.3 Load Balancing	28
	3.3.1 Traditional Solutions	29
	3.3.2 Machine Learning-Based Solutions	29
	3.3.3 Discussion and Insights	30
	3.4 Resource Management	31
	3.4.1 Task-Oriented Resource Management	36
	3.4.2 Virtual Entities-Oriented Resource Management	37

	3.4.3	QoS-Oriented Resource Management	37
	3.4.4	Resource Prediction-Oriented Resource Management	38
	3.4.5	Resource Utilization-Oriented Resource Management	38
	3.4.6	Discussion and Insights	39
3.5	Energy Management		46
	3.5.1	Server Level	47
	3.5.2	Network Level	47
	3.5.3	Data Center Level	48
	3.5.4	Discussion and Insights	49
3.6	Routing Optimization		54
	3.6.1	Intra-DC Routing Optimization	55
	3.6.2	Inter-DC Routing Optimization	55
	3.6.3	Discussion and Insights	55
3.7	Congestion Control		61
	3.7.1	Centralized Congestion Control	62
	3.7.2	Distributed Congestion Control	63
	3.7.3	Discussion and Insights	63
3.8	Fault Management		64
	3.8.1	Fault Prediction	69
	3.8.2	Fault Detection	70
	3.8.3	Fault Location	70
	3.8.4	Fault Self-Healing	70
	3.8.5	Discussion and Insights	71
3.9	Network Security		75
3.10	New Intelligent Networking Concepts		79
	3.10.1	Intent-Driven Network	80
	3.10.2	Knowledge-Defined Network	80
	3.10.3	Self-Driving Network	81
	3.10.4	Intent-Based Network (Gartner)	82
	3.10.5	Intent-Based Network (Cisco)	83
References			84
4	**Insights, Challenges and Opportunities**		101
4.1	Industry Standards		105
	4.1.1	Network Intelligence Quantification Standards	105
	4.1.2	Data Quality Assessment Standards	105
4.2	Model Design		105
	4.2.1	Intelligent Resource Allocation Mechanism	105
	4.2.2	Inter-DC Intelligent Collaborative Optimization Mechanism	106
	4.2.3	Adaptive Feature Engineering	106
	4.2.4	Intelligent Model Selection Mechanism	106
4.3	Network Transmission		106
4.4	Network Visualization		107
References			108

5 Conclusion 109

Index 111

Acronyms

AIMD	Additive Increase Multiplicative Decrease
ANN	Artificial Neural Network
AR	Augmented Reality
ARIMA	Autoregressive Integrated Moving Average Model
ARMA	Autoregressive Moving Average Model
BGP	Border Gateway Protocol
BI	Blocking Island
BMP	BGP Monitoring Protocol
Bof	Bag of Flow
CC	Congestion Control
CFD	Computational Fluid Dynamics
CNN	Convolutional Neural Networks
CRAC	Computer Room Air Conditioner
CRE	Cognitive Routing Engine
DBN	Deep Belief Network
DC	Data Center
DCN	Data Center Network
DDoS	Distributed Denial of Service Attack
DDPG	Deep Deterministic Policy Gradient
DL	Deep Learning
DNN	Deep Neural Network
DPI	Deep Packet Inspection
DQN	Deep Q-Network
DRL	Deep Reinforcement Learning
DT	Decision Tree
ELM	Extreme Learning Machine
eMBB	Enhanced Mobile Broadband
eMDI	Enhanced Media Delivery Index
eMTC	Enhanced Machine-Type Communication
FCT	Flow Completion Time
FPGA	Field Programmable Gate Array

FTR	Fundamental Theory Research
GBDT	Gradient Boosting Decision Tree
GNN	Graph Neural Network
GRU	Gated Recurrent Unit
GWO	Grey Wolf Optimization
HSMM	Hidden Semi-markov Model
IANA	Internet Assigned Numbers Authority
IDS	Intrusion Detection System
IT	Information Technology
LA	Automatic Learning
LSTM	Long Short-Term Memory
MA	Moving Average
MAE	Mean Absolute Error
MAPE	Mean Absolute Percentage Error
ME	Mean Error
ML	Machine Learning
MSE	Mean Squared Error
NBD	Naïve Bayes discretization
NFV	Network Functions Virtualization
NIDS	Network Intrusion Detection System
NMSE	Normalized Mean Square Error
NN	Neural Network
O&M	Operations and Maintenance
PAC	Packaged Air Conditioner
PC	Personal Computer
QoS	Quality of Service
RAE	Relative Absolute Error
RF	Random Forest
RFR	Random Forest Regression
RL	Reinforcement Learning
RMSE	Root Mean Squared Error
RNN	Recurrent Neural Network
RRMSE	Relative Root Mean Squared Error
RRSE	Relative Root Squared Error
RSNE	Ratio of Saved Number of Entries
RTT	Round-Trip Time
SDN	Software Defined Network
SL	Supervised Learning
SLA	Service-Level Agreement
SNMP	Simple Network Management Protocol
SVM	Support Vector Machine
SVR	Support Vector Regression
TCP	Transmission Control Protocol
TPU	Tensor Processing Unit
TWAMP	Two-Way Active Measurement Protocol

UL	Unsupervised Learning
uRLLC	Ultra Reliable Low Latency Communication
VC	Virtual Container
VM	Virtual Machine
VN	Virtual Network
VR	Virtual Reality

Chapter 1
Introduction

As the storage and computation progressively migrate to the cloud, the data center (DC) as the core infrastructure of cloud computing provides vital technical and platform support for enterprise and cloud services. However, with the rapid rise of the data center scale, the network optimization, resource management, operation and maintenance, and data center security have become more and more complicated and challenging. What's more, the burgeoning development of 5G has spawned numerous complex, real-time, diversified, and heterogeneous service scenarios [4, 9], such as enhanced mobile broadband (eMBB) (e.g., ultra-high definition adaptation, augmented reality, virtual reality), ultra reliable low latency communication (uRLLC) (e.g., internet of vehicles, industrial automation, mission-critical applications), and enhanced machine-type communication (eMTC) (e.g., Internet of Things, smart grid, smart cities). The emergence of these new services poses new standards and higher requirements for data centers [28, 32], such as high concurrency, low latency, and micro-burst tolerance.

In terms of data center network (DCN) automation, benefiting from software defined networks (SDNs), data centers have initially achieved automation in some areas, such as automated installation of network policies and automated network monitoring. However, the implementation of such automation typically depends on predefined policies. Whenever the predefined policies are exceeded, the system lacks adaptive processing capability through autonomous learning, and human intervention must be involved. In the face of these challenges and issues, the traditional solutions [1, 5, 23, 35, 37, 39, 44] have become inefficient and incompetent. Moreover, with data availability and security at stake, the issues of data centers are more critical and challenging than ever before [45, 46]. Driven by these factors, about the last decade both academia and industry have conducted extensive research in improving the intelligence level of DCNs by leveraging machine learning (ML) techniques [3, 18, 19, 21, 29–31].

It is universally acknowledged that data-driven ML technologies have made tremendous progress around the last decade [12, 42, 47]. A quantity of academic

© The Author(s), under exclusive license to Springer Nature Singapore Pte Ltd. 2023
T. Wang et al., *Machine Learning Empowered Intelligent Data Center Networking*,
SpringerBriefs in Computer Science, https://doi.org/10.1007/978-981-19-7395-6_1

research has primarily demonstrated that ML could make more effective decisions and optimizations in the ever-changing network environment. With ML technologies, the vast amount of data accumulated in the network can be well exploited to assist the system in dealing with the complex network problems. The substantial increase in computer storage and processing power (e.g., graphics processing unit (GPU) [7], tensor processing unit (TPU) [8]) also provides a strong guarantee for ML implementation in DCNs. The introduction of ML technology will greatly help the data center network to improve the network service quality and the efficiency of operation and maintenance (O&M), so as to cope with the new challenges brought by the increasingly complex network management and dynamic flexible services. With regards to this, various standardization organizations, industries, open-source organizations, and equipment vendors have begun to invest and practice in ML-assisted intelligent data center networking. International standardization organizations such as CCSA and ETSI have started relevant research projects [10, 43]. Open-source organizations such as the Linux Foundation have released several network intelligence related open-source projects. The major operators and equipment vendors have increased their investments and research efforts in network intelligence, and put forward a series of new intelligent networking concepts, such as Juniper's Self-Driving Network [36], Gartner's Intent-Based Network System (IBNS) [20], Cisco's Intent-Based Network (IBN) [16] and Huawei's Intent-Driven Network (IDN) [17].

Although the research on intelligent data center networking has made great progress, it is still confronted with many challenges. On the one hand, the strategy of data collection and processing plays an important role in the effectiveness of data-driven ML-based models. In particular, the way of data collection, the impact of the traffic and computation overhead caused by data collection, and the potential for data leakage are essentially critical. On the other hand, the research on intelligent data center networking is still in the initial stage, where limited by various factors and constraints, the intelligent solutions in some fields are not mature yet, and some intelligence processes are not complete as well. For example, flow prediction plays a crucial role in DCN optimization, which servers as a priori knowledge in routing optimization, resource allocation and congestion control. It can grasp the characteristics and trends of network flow in advance, providing necessary support for relevant service optimization and decision-making. Nevertheless, the huge scale of network and the diversity of services impose great challenges in dealing with such flows with irregular and random distributions in both time and space dimensions. Flow classification, like flow prediction, is widely used as a priori knowledge for a variety of optimization modules, including flow scheduling, load balancing, and energy management. Regarding the quality of service (QoS), dynamic access control, and resource intelligent optimization, accurate categorization of service flows is critical. According to our research (as shown in Chap. 3), the current ML-based traffic classification schemes also have much room for improvement in the fineness of granularity, time efficiency, and robustness. Meanwhile, the goal of load balancing is to guarantee a balanced distribution of flows over multiple network routing paths in order to reduce latency, enhance bandwidth usage, and minimize flow completion time. The problem of load balancing is commonly stated as a multi-commodity flow (MCF) problem, which has been proven to be NP-hard. Undoubtedly, the highly dynamic data center (DC) traffic brings great challenges to the load balancing of intra-DC or inter-DC, which requires an efficient grasp of the characteristics of network traffic. Simultaneously, resource management, as one of the most important optimization problems in the data center, involves the allocation, scheduling, and optimization of computing, storage, network, and other resources, which has a direct impact on the data center's overall resource utilization efficiency and resource availability, as well as the user experience and revenue of service providers. However, with the increasing complexity of network infrastructure, the explosive growth of the number of hardware devices, and the growing demand for services, the traditional unintelligent solutions can no longer effectively deal with these problems, and there is an urgent need for some intelligent resource management solutions. Homoplastically, routing optimization is also one of the most important research areas and has aroused some discussions in both academia and industry. Routing optimization can benefit from SDN by getting a global view of the network and conveniently deploying techniques, however typical SDN-based solutions cannot sensitively react to real-time traffic changes in data center networks. [2, 15, 22, 38, 40, 41]. Notably, the resource management and routing optimization should fully consider the diversity of resources and service requirements, whose multi-objective optimization is usually an intractable problem. Last but not least, the congestion control mechanism of the network also needs further research in terms of model stability, convergence speed, and robustness. The complexity and diversity of service scenarios and finer granularity of flow demands have made congestion control more complicated in data centers. For instance, some applications require high micro-burst tolerance [33, 34], while some applications demand low latency [27] or high throughput [14]. Besides, the diverse applications and computing frameworks with different characteristics in data centers further produce a variety of traffic patterns, such as one-to-one, one-to-many, many-to-one, many-to-many, and all-to-all traffic patterns. However, the traditional transmission control protocol-based (TCP-based) solutions are struggle to match all of these diverse traffic patterns' requirements at the same time [11, 13], resulting in queuing delays, jitter incast, throughput collapse, longer flow completion times, and packet loss [6, 24, 25]. Above all, the high networking complexity, highly dynamic environment, diverse traffic patterns and diversified services all may make it not so easy to directly employ ML techniques to data center. Here we summarize four key challenges encountered in current research, as follows.

- **Data processing:** The ability of data processing and feature engineering directly impacts the performance of ML algorithms. However, the massive volume of real-time data generated in data centers poses a significant challenge to data processing.
- **ML model selection:** The optimization tasks in data centers are complex and diverse, whereas there is no one-size-fits-all ML model than can efficiently deal with all scenarios. Therefore, how to choose the appropriate ML algorithm for

different scenarios and different optimization tasks is a necessary but challenging thing.

- **Collaborative optimization:** Currently, the existing intelligent data center networking solutions usually follow the principle of "one model for one problem". However, the optimization tasks in data centers are numerous with different objectives. Thus, for the scenario of multi optimization tasks, how to achieve an efficient collaborative optimization among multiple intelligent models is a challenging problem.
- **Standardization:** The industry and academia are eagerly waiting for a universally applicable implementation standard to promote DCN intelligence, as many intelligent standards, such as Knowledge-Defined Network (KDN) [26], have not yet been prototyped.

In this survey, we comprehensively investigate the research progress of ML-based intelligent data center networking. Figure 1.1 shows the organization of this book. We classify the existing research work into nine different fields, namely, flow prediction, flow classification, load balancing, resource management, energy

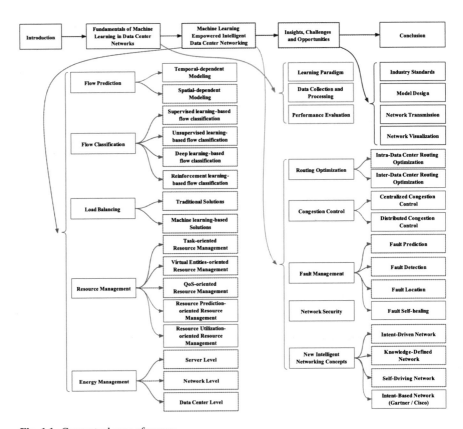

Fig. 1.1 Conceptual map of survey

management, routing optimization, congestion control, fault management, and network security. These existing intelligent DCN solutions in each network area will be analyzed and compared from different dimensions. Furthermore, in-depth insights into the current challenges and future opportunities of ML-assisted DCN will be provided subsequently. The main contributions of this book are summarized as follows:

1. To the best of the authors' knowledge, this is the first comprehensive survey about the application of ML in DCNs. We review the peer-reviewed literature published in recent decade, which are of great influence and well received by peers. The diversity of techniques of machine learning is fully respected to ensure a strong support for the subsequent fair comparisons.
2. We provide enlightening discussions on the usage of ML algorithms in DCNs. We analyze the effectiveness of ML technologies in DCNs from different aspects. In order to provide a multi-dimensional and multi-perspective comparison of various solutions, we innovatively propose our REBEL-3S quality assessment criteria.
3. We extend the study to introduce some new intelligent networking concepts proposed by the leading high tech companies (e.g., Huawei, Cisco, and Juniper), which provides a broad vision of possible research directions in AI-assisted intelligent DCN.
4. We identify a number of research challenges, directions and opportunities corresponding to the open or partially solved problems in the current literature.

The rest of this book is organized as follows. First, we briefly introduce some background knowledge about ML and DCNs in Chap. 2. Then we discuss the wide range of applications of ML in DCNs and provide a comparative analysis from different aspects in Chap. 3. In Chap. 4, we provide insights into DCN's intelligence accompanying by some challenges as well as opportunities. Finally, the book concludes in Chap. 5.

References

1. M. Alizadeh, S. Yang, M. Sharif, S. Katti, N. McKeown, B. Prabhakar, S. Shenker, Pfabric: minimal near-optimal datacenter transport. ACM SIGCOMM Comput. Commun. Rev. **43**(4), 435–446 (2013)
2. F. Amezquita-Suarez, F. Estrada-Solano, N.L. da Fonseca, O.M.C. Rendon, An efficient mice flow routing algorithm for data centers based on software-defined networking, in *ICC 2019-2019 IEEE International Conference on Communications (ICC)* (IEEE, Piscataway, 2019), pp. 1–6
3. J.D.M. Bezerra, A.J. Pinheiro, C.P. de Souza, D.R. Campelo, Performance evaluation of elephant flow predictors in data center networking. Future Gen. Comput. Syst. **102**, 952–964 (2020)
4. L. Chiaraviglio, A.S. Cacciapuoti, G. Di Martino, M. Fiore, M. Montesano, D. Trucchi, N.B. Melazzi, Planning 5G networks under EMF constraints: state of the art and vision. IEEE Access **6**, 51021–51037 (2018)

5. C.H. Chiu, D.K. Singh, Q. Wang, S.J. Park, Coflourish: an sdn-assisted coflow scheduling framework for clouds, in *2017 IEEE 10th International Conference on Cloud Computing (CLOUD)* (IEEE, Piscataway, 2017), pp. 1–8

6. A.K. Choudhury, E.L. Hahne, Dynamic queue length thresholds for shared-memory packet switches. IEEE/ACM Trans. Netw. **6**(2), 130–140 (1998)

7. Cloud gpus (graphics processing units) (2016). https://cloud.google.com/gpu

8. Cloud TPU (2017). https://cloud.google.com/tpu

9. Conformance Specification Radio UEU, 3rd generation partnership project; technical specification group radio access network; evolved universal terrestrial radio access (e-utra); user equipment (UE) conformance specification radio transmission and reception (2011)

10. S. Dahmen-Lhuissier, ETSI - experiential networked intelligence (ENI) (2018). https://www.etsi.org/technologies/experiential-networked-intelligence

11. M. Dong, T. Meng, D. Zarchy, E. Arslan, Y. Gilad, B. Godfrey, M. Schapira, Pcc vivace: online-learning congestion control, in *15th USENIX Symposium on Networked Systems Design and Implementation (NSDI'18)* (2018), pp. 343–356

12. Z.M. Fadlullah, F. Tang, B. Mao, N. Kato, O. Akashi, T. Inoue, K. Mizutani, State-of-the-art deep learning: evolving machine intelligence toward tomorrow's intelligent network traffic control systems. IEEE Commun. Surv. Tuts. **19**(4), 2432–2455 (2017). https://doi.org/10.1109/COMST.2017.2707140

13. T. Flach, N. Dukkipati, A. Terzis, B. Raghavan, N. Cardwell, Y. Cheng, A. Jain, S. Hao, E. Katz-Bassett, R. Govindan, Reducing web latency: the virtue of gentle aggression, in *Proceedings of the ACM SIGCOMM 2013 Conference on SIGCOMM* (2013), pp. 159–170

14. C. Gao, V.C. Lee, K. Li, DemePro: decouple packet marking from enqueuing for multiple services with proactive congestion control. IEEE Trans. Cloud Comput. **99**, 1–14 (2017)

15. Z. Guo, Y. Xu, R. Liu, A. Gushchin, K.y. Chen, A. Walid, H.J. Chao, Balancing flow table occupancy and link utilization in software-defined networks. Future Gen. Comput. Syst. **89**, 213–223 (2018)

16. Intent-based Networking (2017). https://blogs.gartner.com/andrew-lerner/2017/02/07/intent-based-networking/

17. Intent-driven network (2018). https://e.huawei.com/uk/solutions/enterprise-networks/intelligent-ip-networks

18. W. Iqbal, J.L. Berral, D. Carrera, et al., Adaptive sliding windows for improved estimation of data center resource utilization. Future Gen. Comput. Syst. **104**, 212–224 (2020)

19. K. Lei, M. Qin, B. Bai, G. Zhang, M. Yang, Gcn-gan: A non-linear temporal link prediction model for weighted dynamic networks, in *IEEE INFOCOM 2019-IEEE Conference on Computer Communications* (IEEE, Piscataway, 2019), pp. 388–396

20. A. Lerner, J. Skorupa, S. Ganguli, Innovation insight: intent-based networking systems. Tech. Rep., Gartner, Tech. Rep. (2017)

21. B. Li, T. Wang, P. Yang, M. Chen, M. Hamdi, Rethinking data center networks: machine learning enables network intelligence. J. Commun. Inf. Netw. **7**(2), 157–169 (2022)

22. J. Liu, J. Li, G. Shou, Y. Hu, Z. Guo, W. Dai, Sdn based load balancing mechanism for elephant flow in data center networks, in *2014 International Symposium on Wireless Personal Multimedia Communications (WPMC)* (IEEE, Piscataway, 2014), pp. 486–490

23. S. Liu, J. Huang, Y. Zhou, J. Wang, T. He, Task-aware tcp in data center networks. IEEE/ACM Trans. Netw. **27**(1), 389–404 (2019)

24. Y. Lu, X. Fan, L. Qian, Dynamic ECN marking threshold algorithm for TCP congestion control in data center networks. Comput. Commun. **129**, 197–208 (2018)

25. A. Majidi, X. Gao, S. Zhu, N. Jahanbakhsh, G. Chen, Adaptive routing reconfigurations to minimize flow cost in SDN-based data center networks, in *Proceedings of the 48th International Conference on Parallel Processing* (2019), pp. 1–10

26. A. Mestres, A. Rodriguez-Natal, J. Carner, P. Barlet-Ros, E. Alarcón, M. Solé, V. Muntés-Mulero, D. Meyer, S. Barkai, M.J. Hibbett, et al., Knowledge-defined networking. ACM SIGCOMM Comput. Commun. Rev. **47**(3), 2–10 (2017)

27. R. Mittal, V.T. Lam, N. Dukkipati, E. Blem, H. Wassel, M. Ghobadi, A. Vahdat, Y. Wang, D. Wetherall, D. Zats, Timely: RTT-based congestion control for the datacenter. ACM SIGCOMM Comput. Commun. Rev. **45**(4), 537–550 (2015)
28. J. Navarro-Ortiz, P. Romero-Diaz, S. Sendra, P. Ameigeiras, J.J. Ramos-Munoz, J.M. Lopez-Soler, A survey on 5G usage scenarios and traffic models. IEEE Commun. Surv. Tuts. **22**(2), 905–929 (2020)
29. F. Ruffy, M. Przystupa, I. Beschastnikh, Iroko: a framework to prototype reinforcement learning for data center traffic control (2018). arXiv:181209975
30. M.A.S. Saber, M. Ghorbani, A. Bayati, K.K. Nguyen, M. Cheriet, Online data center traffic classification based on inter-flow correlations. IEEE Access **8**, 60401–60416 (2020)
31. T. Scherer, J. Xue, F. Yan, R. Birke, L.Y. Chen, E. Smirni, Practise–demonstrating a neural network based framework for robust prediction of data center workload, in *2015 IEEE/ACM 8th International Conference on Utility and Cloud Computing (UCC)* (IEEE, Piscataway, 2015), pp. 402–403
32. Series M, IMT vision–framework and overall objectives of the future development of IMT for 2020 and beyond. Recommendation ITU 2083 (2015)
33. D. Shan, F. Ren, Improving ECN marking scheme with micro-burst traffic in data center networks, in IEEE INFOCOM 2017-IEEE Conference on Computer Communications (IEEE, Piscataway, 2017), pp. 1–9
34. D. Shan, F. Ren, P. Cheng, R. Shu, C. Guo, Micro-burst in data centers: observations, analysis, and mitigations, in 2018 IEEE 26th International Conference on Network Protocols (ICNP) (IEEE, Piscataway, 2018), pp. 88–98
35. A. Shieh, S. Kandula, A.G. Greenberg, C. Kim, B. Saha, Sharing the data center network, in *8th USENIX Symposium on Networked Systems Design and Implementation NSDI*, vol. 11 (2011), pp. 23–23
36. The self-driving network: sustainable infrastructure (2017). https://www.juniper.net/uk/en/dm/the-self-driving-network/
37. B. Vamanan, J. Hasan, T. Vijaykumar, Deadline-aware datacenter TCP (d2tcp). ACM SIG-COMM Comput. Commun. Rev. **42**(4), 115–126 (2012)
38. Y.C. Wang, S.Y. You, An efficient route management framework for load balance and overhead reduction in SDN-based data center networks. IEEE Trans. Netw. Service Manage. **15**(4), 1422–1434 (2018)
39. T. Wang, Z. Su, Y. Xia, M. Hamdi, Rethinking the data center networking: architecture, network protocols, and resource sharing. IEEE Access, **2**, 1481–1496 (2014)
40. W. Wang, Y. Sun, K. Zheng, M.A. Kaafar, D. Li, Z. Li, Freeway: adaptively isolating the elephant and mice flows on different transmission paths, in *2014 IEEE 22nd International Conference on Network Protocols* (IEEE, Piscataway, 2014), pp. 362–367
41. G. Xiao, W. Wenjun, Z. Jiaming, F. Chao, Z. Yanhua, An openflow based dynamic traffic scheduling strategy for load balancing, in *2017 3rd IEEE International Conference on Computer and Communications (ICCC)* (IEEE, Piscataway, 2017), pp. 531–535
42. J. Xie, F.R. Yu, T. Huang, R. Xie, J. Liu, C. Wang, Y. Liu, A survey of machine learning techniques applied to software defined networking (SDN): research issues and challenges. IEEE Commun. Surv. Tuts. **21**(1), 393–430 (2018)
43. Q. Yang, Y. Liu, T. Chen, Y. Tong, Federated machine learning: concept and applications. ACM Trans. Intell. Syst. Technol. **10**(2), 1–19 (2019)
44. Z. Yao, Y. Wang, J. Ba, J. Zong, S. Feng, Z. Wu, Deadline-aware and energy-efficient dynamic flow scheduling in data center network, in *2017 13th International Conference on Network and Service Management (CNSM)* (IEEE, Piscataway, 2017), pp. 1–4
45. T. Zhang, J. Huang, J. Wang, J. Chen, Y. Pan, G. Min, Designing fast and friendly TCP to fit high speed data center networks, in 2018 IEEE 38th International Conference on Distributed Computing Systems (ICDCS) (IEEE, Piscataway, 2018), pp. 43–53

46. T. Zhang, J. Huang, K. Chen, J. Wang, J. Chen, Y. Pan, G. Min, Rethinking fast and friendly transport in data center networks. IEEE/ACM Trans. Netw. **28**(5), 2364–2377 (2020)
47. Y. Zhao, Y. Li, X. Zhang, G. Geng, W. Zhang, Y. Sun, A survey of networking applications applying the software defined networking concept based on machine learning. IEEE Access **7**, 95397–95417 (2019). https://doi.org/10.1109/ACCESS.2019.2928564

Chapter 2
Fundamentals of Machine Learning in Data Center Networks

In this chapter, we will briefly review the common learning paradigms of ML and some preliminary knowledge about data collection and processing. Furthermore, to better assess the strengths and weaknesses of the existing research work, we design a multi-dimensional and multi-perspective quality assessment criteria, called REBEL-3S.

2.1 Learning Paradigm

Machine learning paradigms can be generally classified into three categories: supervised learning, unsupervised learning, and reinforcement learning. With the in-depth research and development of ML, some new learning paradigms such as deep learning and deep reinforcement learning have been derived for more complex scenarios.

Supervised learning is a simple and efficient learning paradigm, but it requires data to be labeled, where the manual labeling task is usually complex and time-consuming with a considerable workload. This learning paradigm is mainly used to finish simple tasks, e.g., fault prediction and flow classification. The typical representative supervised models are random forests, SVM, KNN, and decision trees. Unsupervised learning can mine potential hidden structures in unlabeled datasets but are relatively fragile and sensitive to data quality. It is more prone to the interference of anomalous data, and the final learning effect is difficult to quantify. Therefore, few research works applied unsupervised learning in DCNs.

Compared with the former two learning paradigms, deep learning is differentiated and characterized by the depth of learning, where it is expected to find the intrinsic association among data through continuous iterative feature extraction, convolution, pooling and other necessary operations. CNN, RNN, LSTM, and GRU are the most common deep learning models used in data center networks. Although accompanied by long training time and slow convergence rate which limit its applicability in some scenarios with high real-time requirements, they are widely used owing to their excellent performance [3, 5, 6, 13].

Unlike the previous models, reinforcement learning is designed to self-learn through continuous interactions with the external environment. The actions performed by the agent can adjust themselves according to the feedback (reward or punishment) given by the environment so as to achieve the global optimal effect. In view of the strong adaptive self-learning ability of reinforcement learning, it has been broadly applied to solve complex problems such as congestion control, routing optimization, and flow scheduling. However, reinforcement learning also has its own problems: (1) The learning model tends to fall into local optimal solutions. (2) The learning model usually requires a long training time, challenging to meet the need of real-time requirements. (3) The learning results may have overfitting phenomenon, which will result in poor model generalization ability in the face of new complex environment.

Deep learning has strong perception ability, but it lacks certain decision-making ability, while reinforcement learning has decision-making ability, yet it has nothing to do with perception problems. Therefore, deep reinforcement learning is proposed by combining the complementary advantages of two learning paradigms, to enable the model with both perceptual ability and decision-making ability. Unfortunately, deep reinforcement learning still retains the problems of poor stability and complex reward functions.

It can be seen that different learning paradigms have specific limitations, and how to adaptively select appropriate ML models for complex optimization tasks with different objectives in data centers is extremely vital and is much difficult.

2.2   Data Collection and Processing

Data collection and processing is regarded as the first step to realize the intelligence of ML-assisted data centers, while the quality of source data directly determines the performance of ML models. However, in the complex and dynamic data center network environment, the massive data generated in real time are usually transient, multidimensional, heterogeneous and diversified, which brings great challenges to the data collection and processing. In this section, we will present our investigation findings and analysis of data collection and processing in data centers from three aspects, namely, data collection scenarios, data collection techniques, and feature engineering. Finally, we will provide our insights into the open problems and challenges in this field.

2.2.1   Data Collection Scenarios

Data collection scenarios in DCNs can be divided into four categories: service data collection, protocol data collection, network performance data collection, and basic network data collection. The service data includes the information of service SLA, and service topology. The service SLA can be further divided into flow-level SLA and service-level SLA. Flow-level SLA measurement is mainly conducted through IFIT (In-situ Flow Information Telemetry) and eMDI. Service-level SLA data can be collected through TWAMP/Y.1731. The protocol data includes protocol stack states, routing information, and delay statistics. The network performance data collection typically consists of interface statistics, queue statistics, and network element health data. The basic network data collection mainly gathers the information of the physical topology, alarms, and logs.

2.2.2   Data Collection Techniques

In the dynamic data center environment, the decision-making of network optimization strategies has strict requirements on the timeliness and quality of the collected data, which also poses a great challenge to the data collection techniques. Empirically, different scenarios have different quality requirements of data collection, thus different data collection techniques will be adopted, accordingly. Generally, the data collection techniques can be grouped into three categories: real time data collection, protocol data collection, and basic data collection. The real-time data collection techniques mainly include Telemetry, IFA, Netflow, sFlow and OpenFlow, which collect the data with a time granularity of seconds reflecting the real time network status. The protocol data collection techniques target at collecting the routing protocol (e.g., BGP) data as well as the topology information, where the typical representatives are BGP-LS and BMP. The SNMP and Syslog are usually employed as the basic data collection techniques to provide basic network information, such as network log, and alarm.

2.2.3   Feature Engineering

Feature engineering refers to the process of transforming the original raw data into the training data, and it directly determines the effectiveness of one ML model, where it can reduce the dimension of data lowering the computing cost. Its essential goal is to improve the performance of the model by acquiring better features of the training data. The feature engineering actions primarily include feature selection and feature extraction. Comparatively, the feature selection plays a more vital role, where high-quality feature selection is helpful to remove redundant irrelevant features and improve the accuracy of ML models. Similar to the backbone internet scenario [1, 4, 9], in DCN scenario, feature engineering can be divided into three different granularity levels: packet-level, flow-level, and application stream-level.

The most fine-grained packet-level features collect packet information and statistics. The flow-level features are generally represented as a 5-tuple, i.e. <source IP address, source port number, destination IP address, destination port number, transport layer protocol>, and flows at this level are usually classified according to the transport layer protocol. The application stream-level is characterized by the number of flows in the Bag of Flow (BoF) level [11, 12], which can be represented as a 3-tuple, i.e., <source IP address, destination IP address, transport layer protocol>. It is suitable for studying the long-term flow statistics of the backbone network at a coarser granularity, but the collection of such high-quality data can increase the computational overhead of the data center.

Feature selection should be adapted to service scenarios. However, in data centers, feature selection strategies are usually not deterministic or invariant. Even for the same problem in the same scenario, features may be inconsistent across different solutions. For example, on the issue of reducing the energy consumption of the data center, Sun et al. [7] simply took the temperature of the chassis as the feature input, while Yi et al. [10] considered the interaction between the average utilization, temperature, and energy consumption of the processor, further expanding the number of key features and effectively exploring the relationship between the features. In conclusion, there are various strategies for feature selection, and the differentiated features will have a direct impact on the final results of the model.

2.2.4 Challenges and Insights

Through the above investigation and analysis, we have summarized the following challenges in current data collection and processing.

- **Data collection load:** The volume of data in DCN has been growing explosively in recent years [8], which already reached 403 exabytes in 2021 [2]. Such massive data bring many problems to data acquisition. For example, whether collecting large volume of data will cause congestion within the collection devices, and whether the data with critical features can be accurately collected. When the data are distributed on different paths, the same data will be collected multiple times on different links if collection points are deployed on all links. Thus, effectively reducing the duplicate sampling of data can help relieve the burden and economic overhead of data collection, and data collection should minimize the impact on the network.

- **Data collection methods:** The dynamic, diversified, complex and real-time DCN environment imposes great challenges to the data collection. Although there are many kinds of data collection methods, they cannot be always well aligned with the collection needs. To put it in practical terms, due to technical implementation limitations, Netstream can only analyze the basic 5-tuple information, and changes beyond the IP header cannot be collected and analyzed. At the same time, the visual display of network data collection results deserve more profound research.
- **Data security and privacy:** The security and privacy of the collected data have become the primary focus of data center networks. A series of behaviors such as data desensitization, access control, and leakage prevention are major issues in the current data center networks.

2.3 Performance Evaluation of ML-Based Solutions in DCN

The performance evaluation of a model should fully consider its specific application scenarios. To provide a multi-dimensional and multi-perspective comparison of various intelligent solutions in the DCN scenario, we propose a quality assessment criteria named REBEL-3S, as illustrated in Table 2.1. "R" stands for reliability, which refers to the robustness and availability of a solution, including the capabilities of failure detection, fault tolerance, and self-healing, etc. "E" stands for energy efficiency, which refers to whether the solution considers the energy cost. "B" and "L" represent bandwidth utilization and latency, respectively. "3S" means security, stability, and scalability of the network, i.e., whether to consider security and privacy against anonymous attacks and abnormal flows, whether to consider network fluctuations and seasonal variation of flows, and whether to consider scalability performance of the solution. It will be marked "YES" to indicate that the solution has taken the above evaluation dimensions into account and vice versa with "NO".

Table 2.1 The meaning of the REBEL-3S

Abbreviations	Properties
R	Reliability
E	Energy Efficiency
B	Bandwidth Utilization
L	Latency
3S	Security, Stability, Scalability

References

1. C. Barakat, P. Thiran, G. Iannaccone, C. Diot, P. Owezarski, Modeling internet backbone traffic at the flow level. IEEE Trans. Signal Process. **51**(8), 2111–2124 (2003)
2. Big data storage use in data centers globally 2015-2021 (2018). https://www.statista.com/statistics/638621/worldwide-data-center-storage-used-by-big-data/
3. X. Cao, Y. Zhong, Y. Zhou, J. Wang, C. Zhu, W. Zhang, Interactive temporal recurrent convolution network for traffic prediction in data centers. IEEE Access **6**, 5276–5289 (2017)
4. C. Fraleigh, S. Moon, B. Lyles, C. Cotton, M. Khan, D. Moll, R. Rockell, T. Seely, S.C. Diot, Packet-level traffic measurements from the sprint ip backbone. IEEE Netw. **17**(6), 6–16 (2003)
5. W.X. Liu, J. Cai, Y. Wang, Q.C. Chen, J.Q. Zeng, Fine-grained flow classification using deep learning for software defined data center networks. J. Netw. Comput. Appl. **168**, 102766 (2020)
6. W. Lu, L. Liang, B. Kong, B. Li, Z. Zhu, Ai-assisted knowledge-defined network orchestration for energy-efficient data center networks. IEEE Commun. Mag. **58**(1), 86–92 (2020)
7. P. Sun, Z. Guo, S. Liu, J. Lan, J. Wang, Y. Hu, Smartfct: improving power-efficiency for data center networks with deep reinforcement learning. Comput. Netw. **179**, 107255 (2020)
8. Synergy Research Group R, Microsoft, amazon and google account for over half of today's 600 hyperscale data centers (2021). https://www.srgresearch.com/articles/microsoft-amazon-and-google-account-for-over-half-of-todays-600-hyperscale-data-centers
9. H. Tao, Z. Hui, L. Zhichun, A methodology for analyzing backbone network traffic at stream-level, in *International Conference on Communication Technology Proceedings, ICCT'03*, vol. 1 (IEEE, Piscataway, 2003), pp. 98–102
10. D. Yi, X. Zhou, Y. Wen, R. Tan, Efficient compute-intensive job allocation in data centers via deep reinforcement learning. IEEE Trans. Parallel Distrib. Syst. **31**(6), 1474–1485 (2020)
11. J. Zhang, C. Chen, Y. Xiang, W. Zhou, Classification of correlated internet traffic flows, in *2012 IEEE 11th International Conference on Trust, Security and Privacy in Computing and Communications* (IEEE, Piscataway, 2012), pp. 490–496
12. J. Zhang, C. Chen, Y. Xiang, W. Zhou, Y. Xiang, Internet traffic classification by aggregating correlated naive bayes predictions. IEEE Trans. Inf. Forensics Secur. **8**(1), 5–15 (2012)
13. K. Zhu, G. Shen, Y. Jiang, J. Lv, Q. Li, M. Xu, Differentiated transmission based on traffic classification with deep learning in datacenter, in *2020 IFIP Networking Conference (Networking)* (IEEE, Piscataway, 2020), pp. 599–603

Chapter 3
Machine Learning Empowered Intelligent Data Center Networking

Machine learning has been widely studied and practiced in data center networks, and a large number of achievements have been made. In this chapter, we will review, compare, and discuss the existing work in the following research areas: flow prediction, flow classification, load balancing, resource management, energy management, routing optimization, congestion control, fault management, network security, and new intelligent networking concepts.

3.1 Flow Prediction

Flow prediction plays a crucial role in DCN optimization, and servers as a priori knowledge in routing optimization, resource allocation and congestion control. It can grasp the characteristics and trends of network flow in advance, providing necessary support for relevant service optimization and decision-making. However, the huge scale of network and the diversity of services impose great challenges in dealing with such flows with irregular and random distributions in both time and space dimensions. For instance, the flow estimation methods based on the flow gravity model [304, 305] and network cascade imaging [110, 203] are challenging to cope with a large number of redundant paths among the massive number of servers in data centers.

The current research work can be generally divided into classical statistical models and ML-based prediction models. The classical statistical models usually include autoregressive models (AR), moving average models (MA), autoregressive moving average models (ARMA) [17], and autoregressive synthetic moving average models (ARIMA) [80]. These models cannot cope with high-dimensional and complex nonlinear relationships yet, and their efficiency and performance in complex spaces are fairly limited. ML-based prediction models can be trained based on historical flow data information to find potential logical relationships

© The Author(s), under exclusive license to Springer Nature Singapore Pte Ltd. 2023
T. Wang et al., *Machine Learning Empowered Intelligent Data Center Networking*, SpringerBriefs in Computer Science, https://doi.org/10.1007/978-981-19-7395-6_3

in complex and massive data, explaining the irregular distribution of network flow in time and space. According to the flow's spatial and temporal distribution characteristics, we classify ML-based prediction solutions into temporal-dependent modeling and spatial-dependent modeling. Next, we will discuss and compare the existing representative work of these two schemes from different perspectives, followed by our insights into flow prediction.

3.1.1 Temporal-Dependent Modeling

The temporal-dependent modeling focuses on the temporal dimension inside the data center. Flow forecasting is usually achieved by using one-dimensional time series data. Liu et al. [154] proposed an elephant flow detection mechanism. They first predicted future flow based on dynamical traffic learning (DTL) algorithm and then dynamically adjusted the elephant flow judgment threshold to improve detection accuracy. However, the frequent involvement of the controller causes extra computational and communication overhead. Beyond these, researchers have also made great efforts to optimize the prediction accuracy with a finer granularity. Szostak et al. [242] used supervised learning and deep learning algorithms to predict future flow in dynamic optical networks. They tested six ML classifiers based on three different datasets. Hardegen et al. [107] collected about 100,000 flow data from a university DCN and used deep learning to perform a more fine-grained predictive analysis of the flow. Besides, researchers [6, 156, 172, 269] have also carried on a lot of innovative work on the basic theoretical research of artificial intelligence. However, some of the experiments to verify the effectiveness of these intelligent schemes are not sufficient. For example, Hongsuk et al. [293] only conducted experimental comparisons on the effectiveness with different parameter settings, lacking the cross-sectional comparisons as aforementioned.

3.1.2 Spatial-Dependent Modeling

The spatial-dependent modeling focuses on both temporal and spatial dimensions across data centers. Nearly 67% of these intelligent solutions compare with the classical statistical models and other ML-based prediction models. The spatial-dependent modeling greatly improves the feasibility and accuracy of solutions, but it also increases the complexity and the operational cost of network O&M. According to our investigations, Over 45% of commercial data center traffic prediction schemes adopt spatial-dependent modeling. Li et al. [142] studied flow transmission schemes across data centers, combined wavelet transform technique with a neural network, and used the interpolation filling method to alleviate the monitoring overhead caused by the uneven spatial distribution of data center traffic. Its experiments conducted in Baidu data center showed that the scheme could reduce the prediction error by

5–30%. We also note that the about 70% of intelligent flow prediction solutions used real-world data. Pfülb et al. [194] based on the real-world data obtained from a university data center that had been desensitized and visualized, the authors used deep learning to predict the inter-DC traffic.

3.1.3 Discussion and Insights

The comprehensive comparisons of the existing approaches are detailed in Table 3.1. Some of our insights into flow prediction are as below.

- **Prior knowledge.** ML algorithms such as Support Vector Regression (SVR) [50] and Random Forest Regression (RFR) [126], compared to classical statistical models, can handle high-dimensional data and obtain their nonlinear relationships well. Nevertheless, their performance in exceptionally complex spatio-temporal scenarios is still limited, partially because they require additional expert knowledge support, where the model learns through the features predesigned by the experts. However, these features usually can not fully describe the data's essential properties.
- **Quality of source data.** The performance of flow prediction heavily depends on the quality of source data, with respect to authenticity, validity, diversity, and instantaneity. Not only for flow prediction, the quality of source data also plays a crucial role in other optimization scenarios of intelligent data center network, which we will explain in Sect. 4.1.2.
- **Anti-interference ability.** The network upgrading, transformation and failures typically can cause sudden fluctuation of traffic, and these abnormal data will interfere with the ultimate accuracy of the model. In order to improve the accuracy of traffic prediction, it is suggested to provide an abnormal traffic identification mechanism to identify the abnormal interference data and eliminate them when executing traffic predictions.

3.2 Flow Classification

Similar to flow prediction, flow classification is also widely used as a priori knowledge for many other optimization modules such as flow scheduling, load balancing, and energy management. Accurate classification of service flows is essential for QoS, dynamic access control, and resource intelligent optimization. The daily operation and maintenance also require accurate classification of unknown or malicious flows. Moreover, a reasonable prioritized classification ordering can help enterprise network operators optimize service applications individually and meet the resource management requirements and service needs. Nevertheless,

Table 3.1 Research progress of data center network intelligence: flow prediction

Ref	Category[a]	ML category and basic model	Features	Estimation function		Experimental comparison subjects[b]
Pfülb et al. [194]	S	DL, DNN	Trained with a large dataset of approximately 50 million streams for more granular traffic prediction and visual analysis	Accuracy, etc.	Collected data from the networks at Fulda University of Applied Sciences	✓|✗|✗
Liu et al. [154]	T	FTR, DTL	Adopted dynamical flow learning (DTL) algorithm, weighted optimization based on Gaussian distribution and smooth mechanism based on difference estimation	Customized	CAIDA [46]	✓|✗|✓
Szostak et al. [241, 242]	T	SL/DL, KNN etc.	A traffic prediction method in dynamic optical networks for serving VNF was proposed	Customized	Simulated data	✓|✗|✓
Yi et al. [293]	T	DL, DNN	The first research team to use TensorFlow-based DNN for traffic prediction	Customized	Obtained from about 0.5 million probe vehicles with an on-board device (OBD)	✗|✗|✓
Wang et al. [269]	T	FTR, None	Installed preclassified information into the controller for fast classification and use OpenFlow for event management	None	No experiments	✗|✗|✗
Hardegen et al. [106]	S	DL, DNN	Unlike the binary classification of "mice" and "elephant", the authors triple-classified according to the predicted bit rate, while using pre-processing, anonymization, and visualization techniques	Accuracy, etc.	Data collection took place in a real-world production network at Fulda University of Applied Sciences	✓|✗|✗

Reference	S/T	Method	Description	Metric	Dataset			
Mozo et al. [179]	S	DL, CNN	Multiresolution strategy multiple-channel convolutions to incorporate multiresolution context	MSE/MAE	ONTS [186]	✓	✓	✓
Lei et al. [138]	S	DL, LSTM, etc.	The dynamic prediction of the network was formulated as a temporal link prediction, combined various NN structures, took various evaluation measures, and conducted adequate experiments	Customized	UCSB[205], KAIST[136], BJ-Taxi[300], and NumFabric[182]	✓	✓	✓
Cao et al. [44]	S	DL, GRU/CNN	Learned network flow as images to capture the network-wide services' correlations	RMSE	Historical data from Yahoo's Data Center	✓	✓	✓
Li et al. [142, 143]	S	DL, DNN	Combine wavelet transform with DNN to improve prediction accuracy	RRMSE	A production data center with tens of thousands of servers from Baidu for six weeks	✓	✓	✓
Mbous et al. [172]	T	FTR, None	Kalman filter-based algorithm	Utilization rate, etc.	Simulated data	✓	✓	✗
Shi et al. [229]	T	DL, LSTM	Traffic forecasting in hybrid data center networks to aid with optical path reconfiguration	MSE	Simulated data	✓	✗	✗
Balanici et al. [28–30]	T	DL, DNN	Primary Application Solution	MSE, MAE	Mix of simulated and real-world data	✓	✗	✗
Singh et al. [234]	S	DL, LSTM	A multi-class dynamic service model was considered	RMSE	Simulated data	✓	✗	✗
Estrada-Solano et al. [78]	S	DL, DT	Incremental learning-based network flow prediction	FPR, etc.	University data centers [32]	✗	✓	✓

(continued)

Table 3.1 (continued)

Ref	Category[a]	ML category and basic model	Features	Estimation function		Experimental comparison subjects[b]
Bezerra et al. [37]	S	DL, RNN	Proposed a hybrid prediction model based on FARIMA and RNN models	RMSE, MAPE, etc.	Historical data from Facebook's Data Center	✓|✓|✓
Hardegen et al. [107]	T	DL, DNN	More granular network flow prediction	Accuracy, etc.	University network flow data, about 100,000 records	✓|✗|✗
Yu et al. [298]	T	DL, RNN/LSTM	Achieved four classification based on time and frequency, combined with RNN and LSTM to propose a new flow prediction method	Customized	Simulated data	✓|✓|✓
Liu et al. [156]	T	FTR, Unspecified	Learning multiple historical traffic matrixes (TMs) through gradient boosting machine (GBM) method rather than monitoring each flow	NMSE, RSNE	Real-world data from unknown sources	✓|✓|✗
Luo et al. [165]	S	SL/DL, KNN/LSTM	A hybrid flow prediction methodology, highlights the importance of the highly relevant stations to the prediction result	RMSE, Accuracy	Provided by the Transportation Research Data Lab (TDRL)	✓|✓|✓
Nie et al. [184]	T	DL, DBN	A flow prediction method based on DBN was proposed and learned the statistical properties between flow start and end nodes. A network tomography model was also constructed	Standard deviation, etc.	Real-world network flow data from unknown sources	✗|✓|✓

Reference	S/T	Methods	Description	Decision time etc	Data			
Aibin, Michal et al. [6, 7]	S	FTR, MCTS	An evaluation of a specific provider-centric use case for control and provisioning of DC services in WANs.	Decision time etc	Simulated data	✓	✓	✓
Yu et al. [299]	T	SL/DL, SNN	Proposed a supervised spiking NN (s-SNN) framework with multi-synaptic mechanism and error feedback model for flow prediction in hybrid E/O switching intra-datacenter network	Resource occupation rate, etc.	Packet head information collected every five minutes for 10 days from three university data centers in Beijing, China	✓	✗	✓
Paul et al. [193]	T	DL, LSTM/GRU	Analysed the performances of different RNN models with activation functions to obtain future flow demands.	RMSE, SMAPE	Used part of the data released by Telecom Italia in 2015	✓	✗	✗
Guo et al. [97]	S	DL, DNN	Designed an adaptive and scalable downlink based flow predictor that exploits the temporal and spatial characteristics of inter-DC flows and provides accurate and timely forecasts	SSE	Simulated data	✓	✗	✗
Zhu et al. [312]	T	DL/UL, LSTM/BIRCH	Used LSTM based NN to predict the arrival of jobs and aggregate requests for computing resources	MSE, etc.	ClusterData2011_2 [209], which is a data set released by Google Data Center	✓	✗	✓

[a] For simplicity, "S" indicates Spatial-dependent Modeling and "T" denotes Temporal-dependent Modeling

[b] The notations ✓ and ✗ in this column are used to indicate whether the scheme has carried out the specified experiments. As shown in this column, from left to right, it represents self-comparison with different parameters, comparisons with traditional schemes, and comparisons with intelligent schemes, respectively. Unless otherwise specified, it has the same meaning in subsequent tables

the highly dynamic and differentiated traffic, and complex traffic transmission mechanism greatly increases the difficulty of traffic classification.

Traditional traffic classification schemes are usually based on the information of port, payload, and host behaviors. In the early stages of the Internet, most protocols used well-known port numbers assigned by the Internet Assigned Numbers Authority (IANA) [224]. However, protocols and applications began to use random or dynamic port numbers so as to hide network security tools. Some experimental results further show that port-based classification methods are not very effective, for example, Moore et al. [177] observed that the accuracy of the classification techniques based on IANA port list does not exceed 70%. To overcome the limitations of the above classification techniques, the payload-based flow classification method was proposed as an alternative. The payload-based approach, also known as deep packet inspection (DPI), classifies flows by examining the packet payload and comparing it with the protocols' known signatures [82, 85, 93, 223]. Common DPI tools include L7 filter [19] and OpenDPI [38]. However, such DPI-based solutions incur high computation overhead and storage cost though they can achieve higher accuracy of traffic classification than port-based solutions. Although the accuracy is improved compared to the previous methods, the complexity and computational effort are significantly higher. Furthermore, dealing with the increasingly prominent network privacy and security issues also brings high complexity and difficulty to DPI-based techniques [34, 77]. Thus, some researchers put forward a new kind of flow classification technique based on host behaviors. This technique uses the hosts' inherent behavioral characteristics to classify flows, overcoming the limitations caused by unregistered or misused port numbers and high loads of encrypted packets. Nevertheless, the location of the monitoring system largely determines the accuracy of this method [41], especially when the observed communication patterns may be affected by the asymmetry of routing.

In the face of such dilemma of traditional solutions, ML-based flow classification techniques can address the mentioned limitations effectively [65, 286]. Based on the statistical characteristics of data flows, they complete the complex classification tasks with a lower computational cost. Next, we will introduce and discuss different ML-based flow classification techniques according to the types of machine learning paradigms, and followed by our insights at the end.

3.2.1 Supervised Learning-Based Flow Classification

Supervised learning can achieve higher accuracy of classification among applications. Despite of the tedious labeling work, many supervised learning algorithms have been applied in flow classification, including decision trees, random forests, KNN, and SVM. Trois et al. [255] generated different image textures for different applications, and they classified the flow matrix information using supervised learning algorithms, such as SVM and random forests. Zhao et al. [310] applied supervised learning algorithms to propose a new classification model that achieved an accuracy of about 99% in a large supercomputing center.

3.2.2 Unsupervised Learning-Based Flow Classification

Unsupervised learning-based flow classification techniques do not require labeled datasets, eliminating the difficulties encountered in supervised learning and providing higher robustness. In contrast to supervised learning, the clusters constructed by unsupervised learning need to be mapped to the corresponding applications. However, the large gap between the number of clusters and applications makes it more challenging to classify flows. As investigated in the work of Yan et al. [288], many existing flow classification schemes have adopted unsupervised learning algorithms [24, 35, 76, 270, 307]. Xiao et al. [281] focused on the imbalance characteristics of elephant and mice flows in DCNs and proposed a flow classification method using spectral analysis and clustering algorithms. Saber et al. [214] had a similar research concern and proposed a cost-sensitive classification method that can effectively reduce classification latency. Deque-torres et al. [74] proposed a knowledge-defined networking (KDN) based approach for identifying heavy hitters in data center networks, where the efficient threshold for the heavy hitter detection was determined through clustering analysis. Unfortunately, the scheme was not compared with other intelligent methods, thus failing in proving its superiority.

3.2.3 Deep Learning-Based Flow Classification

The service data and traffic data generated in the data center networks are typically massive, multidimensional, and interrelated. It's very challenging to explore the valuable relationship between these data. To this end, deep learning (e.g., CNN, RNN and LSTM) is introduced to DCN as a promising way to find the potential relationship between these massive and interrelated data. Compared with the former two ML-based classification techniques, the deep learning based schemes have no advantage in training time and classification speed. To this end, Wang et al. [273] focused on the speed of classification and implemented a high-speed online flow classifier via field programmable gate array (FPGA), where the authors claimed that it can guarantee an accuracy of more than 99% while reducing the training time to be one-thousandth of the CPU-based approach. Liu et al. [159] implemented a more fine-grained flow classification method based on GRU and reduced flow monitoring costs. In addition, Zeng et al. [302] proposed a lightweight end-to-end framework for flow classification and intrusion detection by deeply integrating flow classification and network security.

3.2.4 Reinforcement Learning-Based Flow Classification

Reinforcement learning agent iteratively interacts with the environment aiming to find a global optimal classification scheme according to the feedback reward and punishment of feedback in a network scenario. To handle the highly dynamic network conditions in DCNs, Tang et al. [244] proposed a new reinforcement learning-based flow splitter that effectively reduced the average completion time of flows, especially for delay-sensitive mice flows. Whereas, as the reinforcement learning tends to fall into local optimal solution and takes a longer training time, this paradigm has not been widely used in traffic classification.

3.2.5 Discussion and Insights

The ML can overcome the limitations and constraints of traditional flow classification schemes. In view of this, numerous ML-based intelligent traffic classification schemes have been proposed. Table 3.2 summarizes and compares these existing work from various perspectives. Through systematic investigations and in-depth analysis, in this book we summarize a general flow classification workflow, as shown in Fig. 3.1, ranging from different levels of feature collections, data pre-processing, model training, to model inference outputting classification results. Furthermore, we have summarized several key concerns that need to be addressed, as listed below.

- **Fine granularity.** The complex diverse DCN service scenarios, high requirements on flow control, and more precise network management are driving the flow classification techniques toward a more fine-grained direction. A fine-grained and accurate classification scheme can allocate network resources efficiently, ensuring a better user experience. However, most of the traditional and intelligent classification schemes are only based on a rough general classification scale, for example, singly based on the network protocol or a single function of the application, which can not provide better QoS. In some simplified scenarios, even if the fine-grained classification has been achieved, the computational overhead, monitoring overhead, stability and feasibility also need to be considered.
- **Flexibility and robustness.** To meet various service needs, flow classification should consider the timeliness and effectiveness of classification, which could help services meet their SLAs. Using FPGA is a feasible way to improve the response speed of classification and avoid the influence of abnormal conditions on the classification efficiency. When encountering the common network anomalies such as jitter, packet loss, and retransmission, the efficiency of a robust flow classification solution should not degrade. Moreover, the quality of extracted data features can also significantly affect the final result of classification, and redundant features will reduce the accuracy of the algorithm along with additional computational overhead [230].

Table 3.2 Research progress of data center network intelligence: flow classification

Ref	ML category and model adopted	Features	Data source	Additional constraints	Estimation function	Experimental comparison subjects
Zeng et al. [302]	DL, CNN/LSTM	A lightweight end-to-end framework for flow classification and intrusion detection was proposed, and it had excellent performance on two public data sets	ISCX 2012 IDS dataset [231] and ISCX VPN-nonVPN flow dataset [72]	Storage	F_1-score, etc.	✗|✗|✓
Xiao et al. [281]	UL, clustering algorithm	A fast and effective flow classification method was proposed.	Real-world historical data [171]	Statistical cost	Class accuracy and class recall	✗|✗|✓
Liu et al. [154]	FTR, unspecified	Adopted dynamical flow learning (DTL) algorithm, weighted optimization based on Gaussian distribution and smooth mechanism based on difference estimation.	CAIDA [46]	None	Customized	✓|✗|✓
Duque-Torres et al. [74]	UL, clustering algorithm	A new intelligent system based on KDN	Real-world historical data from University DCN	None	Silhouette score	✓|✗|✗
Cuzzocrea et al. [64]	FTR, KNN, etc.	ML classifiers based on different workload models produced the best classifiers	Simulated data	Workload	Accuracy	✓|✗|✓
Shekhawat et al. [227]	FTR, SVM etc.	A method for classifying and characterizing data center workloads based on resource usage was proposed	Google Cluster Trace (GCT) [211] and Bit Brains Trace (BBT) [228]	Workload	SSE	✓|✗|✓

(continued)

Table 3.2 (continued)

Ref	ML category and model adopted	Features	Data source	Additional constraints	Estimation function	Experimental comparison subjects
Zhu et al. [314]	DL, GRU	Differentiated transmission control services	UNB CIC VPN-nonVPN dataset [72]	Buffer size, FCT and throughput	F_1-score etc.	✓|✓|✓
Trois et al. [255]	SL, SVM/RF	Flow forecasting based on flow matrices	gathered from MapReduce [68] and DCNs	None	Confusion matrix etc.	✗|✗|✓
Liu et al. [159]	DL, GRU/RF	Fine-grained approach to flow classification	Real-world historical flow data from a data center [33]	Monitoring cost	F_1-score etc.	✓|✗|✓
Tang et al. [244]	RL, DDPG	Proposed high-performance stream scheduling policy DDPG-FS based on DDPG	Simulated data	Load, FCT	FCT etc.	✓|✓|✓
Viljoen et al. [264]	DL, DNN	A low overhead, adaptive traffic classifier was proposed	Historical flow data from a data center [33]	None	Confusion matrix etc.	✗|✗|✓
Zhu et al. [312]	DL/UL, LSTM/BIRCH	The unsupervised hierarchical clustering algorithm was used to classify the unexecuted jobs, and Davies-Bouldin and other indicators were used to evaluate the clustering quality	ClusterData2011_2 [209], which is a data set released by Google Data Center	Workload	MSE, etc.	✓|✗|✓
Amaral et al. [15]	SL, RF	The traffic data was collected by Openflow protocol and analyzed by machine learning algorithms such as RF	Real-world historical data from unknown sources	None	Accuracy	✓|✗|✓

Reference	Type	Description	Dataset	Network metric	Evaluation metric	
Bolodurina et al. [39]	Unspecified, unspecified	Accelerated learning of the initial phase of analysis performance, improve flow classification accuracy	Simulated data	Network latency	Delay of application request	✓|✗|✓
Wang et al. [273]	DL, DNN	Implementing a flow classifier using FPGAs with far better performance than CPU	Real-world historical data from unknown sources	None	Training time and accuracy	✓|✓|✗
Sarber et al. [214]	SL/UL, BRF/RKNN	A cost-sensitive classification method was proposed to guarantee performance while alleviating the data imbalance problem	CAIDA [46], two university datacenters datasets[248] and an internet flow dataset (UNIBs [258])	Network resources and latency	F_1-score, etc.	✓|✓|✓
Wang et al. [271]	SL, C4.5/NBD	A packet optical switching network architecture and a priority-aware scheduling algorithm were proposed	Simulated data	Bandwidth	Recall and classification speed	✓|✓|✗
Zhao et al. [310]	SL, RF/C4.5/KNN	Proposed a clustering flow label propagation technique based on equivalent flow-labeled propagation and a synthetic-flow feature generation algorithm based on Bidirectional-flow (BDF)	Real-world historical data from unknown sources	None	Precision, recall, F_1-score and accuracy	✓|✗|✓

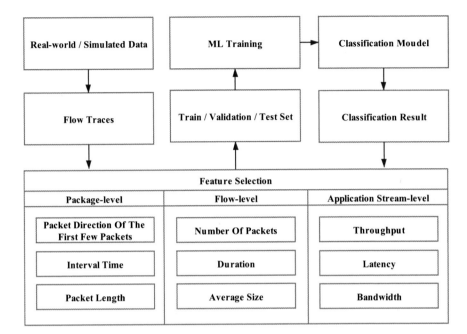

Fig. 3.1 The general system workflow of flow classification, ranging from different levels of feature collections, data pre-processing, model training, to the final model inference outputting classification results

3.3 Load Balancing

The purpose of load balancing is to ensure the balanced distribution of flows on different network routing paths, so as to minimize flow completion time, improve bandwidth utilization and reduce latency. The load balancing problem is usually formulated as a multi-commodity flow (MCF) problem, which has been proved to be NP-hard. The traffic in data centers usually changes in milliseconds or even microseconds, however, traditional unintelligent solutions lack the flexibility of dynamic adjustment according to the real-time network environment status, which may lead to imbalanced load distribution or even network congestion [306]. As for the performance evaluation and effectiveness verification, there are a variety of metrics. Generally speaking, solutions are usually evaluated in terms of the average link utilization, scalability, robustness, and energy efficiency, which is consistent with the evaluation dimension of REBEL-3S.

3.3.1 Traditional Solutions

Empirically, the decision-making of load balancing largely depends on the real-time collected network running status data. According to the way of data acquisition, traditional unintelligent solutions can be divided into two categories: centralized and distributed. Centralized solutions, such as DENS [133], Hedera [9], and Mahout [63] make decisions based on the global network knowledge acquired through a centralized controller. However, the centralized schemes typically inevitably result in additional communication overhead between the controller and the data plane, which poses extra traffic burden on the network. Besides, the centralized schemes usually require dedicated and customized hardware (e.g. OpenFlow supported), which are cross vendor incompatible. Distributed solution is difficult to make the best decision without a global view. Although the topology of DCNs is often symmetrical in design, it is still difficult to deal with the network failure caused by the damage to hardware devices. Alizadeh et al. [14] insisted that an efficient load balancing scheme must address the asymmetry issue caused by network failures which are highly disruptive. Importantly, traditional approaches are difficult to learn from the historical traffic data and automatically adjust the strategies to achieve network optimization.

3.3.2 Machine Learning-Based Solutions

Facing the ever-changing network environment, ML can help the network self-learning, realize the self-decision of flow scheduling strategy, and self-adaptation to the network environment. Zhao et al. [309] proposed two ML-assisted flow aggregation schemes to achieve low latency and high bandwidth. They improved network throughput through specifically designed optical cross-connect switches, and deployed ML algorithms (such as DT, KNN, and SVM) with relaxed accuracy requirements to edge nodes to reduce latency. The Wavelength Division Multiplexing (WDM) technology was used to improve the scalability of the optical network, but the FPGA board was installed on each ToR to perform feature sampling, increasing the hardware cost. Wang et al. [271] used supervised learning algorithms, such as C4.5, to classify network flows with different characteristics and developed a priority-aware scheduling algorithm for packet switching. The simulation experiments showed that their scheduling algorithm was superior to the classical RR algorithm [103] with respect to the average delay and packet loss rates. Compared with the former two kinds of learning paradigms, deep learning algorithms have better applicability and more and more researchers prefer to use them. Li et al. [141] designed a GNN-based optimizer for flow scheduling to reduce the flow completion time (FCT). However, GNN brings a more complex network structure and increases computational costs. Prevost et al. [200] devoted to the energy consumption problem caused by the imbalanced load. They proposed a new DNN-based framework to achieve load demand prediction and stochastic state transfer. With the increasing difficulty of optimization goal, more and more researchers consider using reinforcement learning and deep reinforcement learning to deal with the dynamic network environment. Tang et al. [244] employed a modified DDPG algorithm for high-performance flow scheduling. Compared to the native DDPG and traditional unintelligent methods, their solution significantly reduced the FCT of delay-sensitive flows.

3.3.3 Discussion and Insights

To avoid the difficulty of collecting real-world data, 90% of the intelligent solutions were tested based on simulation-generated data. Due to the diversity and complexity of DCN application scenarios and the differences in data sources and scenarios, it is not easy to make a fair comparison between intelligent schemes. As a result, over 76% of the solutions lack comparisons to other ML-based solutions. Besides, inheriting the advantages of traditional unintelligent solutions, more than 40% of intelligent solutions adopt SDN architecture to collect network data and make decisions based on a centralized controller (Table 3.3).

The details of the existing intelligent solutions are listed in Table 3.4, and the assessment results of each solution based on REBEL-3S are summarized in Table 3.3. Clearly, most solutions consider bandwidth utilization and latency, account for 94 and 88% respectively, while few solutions take security and reliability into account, account for 6 and 12% respectively. In addition to the issues and challenges discussed above, the following two concerns need to be considered.

- **Compatibility of network stacks.** According to the work of Wang et al. [274], one of the necessary conditions for many traditional research work is to be compatible with different network protocol stacks [12, 13, 26, 51, 112, 278]. With the advent of new network protocols such as D^2TCP [262] and DCTCP [11], the unequal distribution of network bandwidth among different network users due to stack incompatibility has attracted considerable attention [62, 109]. Intelligent solutions should focus on compatibility between different network protocols and prevent unfair resource allocation caused by the different protocol parameters.
- **Dynamic of network flow.** In respect of the dynamic of network traffic, it is necessary and beneficial to adjust the threshold and priority of flows in time. It is suggested to pay attention to the local and overall benefits, especially for the mixed flow scheduling problem in the scheduling process.

Table 3.3 Assessment of load balancing schemes based on REBEL-3S

Ref	Reliability	Energy efficiency	Bandwidth utilization	Latency	Security	Stability	Scalability
Ruelas et al. [212]	NO	NO	YES	YES	NO	NO	YES
Tosounidis et al. [253]	NO	NO	YES	YES	NO	YES	YES
Doke et al. [69]	NO	NO	YES	YES	NO	NO	NO
Hashemi et al. [108]	YES	NO	YES	YES	NO	YES	NO
Zhou et al. [311]	NO	NO	YES	YES	NO	YES	NO
Tang et al. [244]	NO	NO	YES	YES	NO	YES	YES
Liu et al. [148, 162]	NO	NO	YES	YES	NO	YES	YES
Yu et al. [297]	YES	NO	YES	YES	NO	YES	YES
Zhao et al. [309]	NO	NO	YES	YES	NO	YES	YES
Liu et al. [158]	NO	NO	YES	NO	NO	YES	YES
Francois et al. [84]	NO	NO	YES	YES	YES	YES	YES
Scherer et al. [220]	NO	NO	YES	YES	NO	NO	NO
Prevost et al. [200]	NO	YES	NO	NO	NO	NO	NO
Sun et al. [239, 240]	NO	YES	YES	YES	NO	YES	YES
Lin et al. [147]	NO	NO	YES	YES	NO	YES	NO
Wang et al. [271]	NO	YES	YES	YES	NO	YES	YES
Li et al. [141]	NO	NO	YES	YES	NO	NO	YES

3.4 Resource Management

As one of the most critical optimization problems in data center, resource management involves the allocation, scheduling, and optimization of computing, storage, network and other resources, which directly affects the overall resource utilization efficiency and resource availability of data center, and further affects the user experience and the revenue of service providers. However, with the increasing complexity of network infrastructure, the explosive growth of the number of hardware devices, and the growing demand for services, the traditional unintelligent solutions can no longer effectively deal with these problems, and there is an urgent need for some intelligent resource management solutions. Studies reveal that ML-assisted intelligent resource allocation can maximize the profit of service providers, provide better quality of experience (QoE) for tenants, and effectively reduce energy costs.

Table 3.4 Research progress of data center network intelligence: load balancing

Ref	ML category and model adopted	Features	Data source	Feature selection	Additional constraints	Estimation function	Experimental comparison subjects
Ruelas et al. [212]	DL, DNN	Based on the ideas of KDN and DNN, a new load balancing method was provided	Simulated data	Bandwidth, Latency	None	MSE	✓|✗|✗
Tosounidis et al. [253]	DRL, DQN	Deep reinforcement learning was used to efficiently load balance service requests in DCNs	Simulated data	Bandwidth/CPU/Memory utilization	CPU computing power	RTT, etc.	✓|✓|✗
Doke et al. [69]	DRL, DQN	Try to apply DQN algorithm to network traffic management	Simulated data	Number of session requests	None	Standard deviation	✓|✓|✗
Hashemi et al. [108]	DL, CNN	An end-to-end real-time flow management system based on deep learning was proposed	Simulated data	Time-varying flow sizes	Accidents and adverse weather conditions, etc.	AvgRMSE, etc.	✓|✓|✗
Zhou et al. [311]	DL, RNN	A residual flow compression mechanism was introduced to minimize the completion time of data-intensive applications	Simulated data	Coflow width, coflow size and arrival time, etc.	Network bandwidth	FCT, etc.	✓|✓|✗
Tang et al. [244]	RL, DDPG	Proposed high-performance stream scheduling policy DDPG-FS based on DDPG	Simulated data	Maximum operational flow between links and maximum total flow demand, etc.	Load	AvgFCT, etc.	✓|✓|✓

Liu et al. [148, 162]	DRL, DQN/DDPG	The DRL-R (Deep Reinforcement Learning-based Routing) algorithm was proposed to bridge multiple resources (node cache, link bandwidth) by quantifying the contribution of multiple resources (node cache, link bandwidth) to reduce latency	Simulated data	Node cache, Link bandwidth	Maximize overall network throughput while meeting QoS	FCT, etc.	✓|✓|✓
Yu et al. [297]	DL, BRNN/LSTM	A flow scheduling method capable of extracting long-term flow characteristics	Historical flow data obtained from three university data centers	5-tuple, etc.	Network Resource Utilization	MAE, MRE, and RMSE, etc.	✓|✗|✓
Zhao et al. [309]	SL/UL, SVM/DT, etc.	Proposed two ML-assisted traffic aggregation schemes that can effectively improve throughput, reduce latency and FCT	Simulated data	Network throughput, network latency and flow completion time	Transaction response processing time	Accuracy, etc.	✓|✓|✗
Liu et al. [158]	DRL, DDPG/CNN	A hybrid flow scheduling scheme based on deep reinforcement learning was proposed	Simulated data	Paths and flows information	Maximizing deadline satisfaction rate for mice flows and minimizing FCT for elephant flows	Deadline meet rate [153]	✓|✓|✗

(continued)

Table 3.4 (continued)

Ref	ML category and model adopted	Features	Data source	Feature selection	Additional constraints	Estimation function	Experimental comparison subjects
Francois et al. [84]	DL/RL, RNN	A logical centralized cognitive routing engine was developed based on stochastic NNs with reinforcement learning	5 geographically dispersed data centers	Latency	None	RTT, etc.	✓|✓|✗
Scherer et al. [220]	DL, DNN	A neural network-based framework for server workload prediction was proposed	VM workload tracking and real-time data from real-time systems recorded in private cloud data centers operated by IBM	CPU, memory, disk, and network	None	Accuracy	✓|✗|✓
Prevost et al. [200]	DL, DNN	A new framework combining load demand forecasting and stochastic state transfer models was proposed to minimize energy consumption while maintaining network performance	URL resource requests for NASA's web server and EPA's web server	Unspecified	Minimize energy consumption while guaranteeing performance	RMSE, RMSE	✓|✓|✗

Sun et al. [239, 240]	DRL, DDPG	The DRL algorithm was used to improve power efficiency and ensure FCT	Wikipedia trace files [259]	Number of switch ports, port power, incoming and outgoing flow rates, etc.	QoS and energy consumption	FCT, etc.	✓	✓	✗
Lin et al. [147]	DRL, DDPG	Reinforcement-based learning was used to learn a network and perform load balancing by aggregating flows	Simulated data	Throughput between nodes	Network Overheads	Customized	✓	✓	✗
Wang et al. [271]	SL, C4.5/NBD	A packet-switched optical network (PSON) architecture and a priority-aware scheduling algorithm were proposed	Simulated data	Virtual queue priority, number of packets, latency, etc.	Bandwidth	Recall and classification speed	✓	✓	✗
Li et al. [141]	DL, GNN	To be able to support relational reasoning and combinatorial generalization, the authors have proposed a GNN-based flow optimization method	Simulated data	5-tuple, bandwidth	None	FCT, etc.	✓	✓	✗

There has been a wide variety of resource management solutions, such as multi-level queues [129], simulated annealing [189], priority-based [236], and heuristic algorithms [36]. The advent of virtualization allows virtual machines (VMs), virtual containers (VCs), and virtual networks (VNs) to be implemented on a shared physical server. Whereas, the association between various hardware resources (such as CPU, memory and disk space) and virtual resources is highly dynamic throughout the life cycle of services, which is difficult to grasp clearly. The preliminary research findings demonstrated that traditional unintelligent resource management methods can not mine the potential relationships between complex parameters quickly and dynamically. Besides, multi-objective optimization also increases the difficulty of network optimization, such as considering QoS, energy cost and performance optimization at the same time. Furthermore, in a large-scale data center, the complex configuration is also a challenging and destructive problem, where once the configuration error occurs, it will cause incalculable damage to network services, especially for latency-sensitive services. ML can make up for the deficiency of traditional unintelligent methods by learning historical data to dynamically make appropriate management strategies adaptively. Therefore, many researchers have begun to study in this direction, and the solution combined with machine learning came into being. The work of Fiala and Joe [81] explored the application of ML techniques for resource management in the cloud computing area, and Murali et al. [180] focused on a distributed NN-based ML approach to achieve efficient resource allocation.

In data center networks, the types of network resources are rich and diverse. At the network level, it can be a physical hardware resource (server, switch, port, link, CPU, memory) or an abstract software resource (virtual network, virtual node, virtual link, virtual switch). In addition, network resources can be task/job-oriented or QoS-oriented. From the perspective of the resource life cycle, resource management can also focus on resource prediction or resource utilization optimization. In view of the diversity of resource management methods and the difference of optimization objectives, we divide ML-based resource management schemes into the following five categories: task-oriented, virtual entity-oriented, QoS-oriented, resource prediction-oriented, and resource utilization-oriented resource management.

3.4.1 Task-Oriented Resource Management

In data centers, there are various special tasks with different particular performance requirements, such as compute-intensive tasks and latency-sensitive tasks, which require that the resource management solutions can be customized for different tasks. Tesauro et al. [247] used reinforcement learning to optimize the allocation of computing resources by global arbitration, allocated efficient server resources (such as bandwidth and memory) for each web application, and solved the limitations of reinforcement learning by queuing model policy. Marahatta et al. [170] classified tasks into failure-prone and failure-prone tasks by DNN and executed different allocation policies for different types of tasks. In addition, resource management can also help reduce the energy consumption of the data center. Yi et al. [294] scheduled the load of computationally intensive tasks based on deep reinforcement learning to minimize the energy cost.

3.4.2 Virtual Entities-Oriented Resource Management

Virtualization allows tasks with different service and performance requirements to share a series of resources. Generally speaking, virtualized entities include virtual machines (VMs), virtual containers (VCs), and virtual networks (VNs), and we define the solutions that allocate resources for virtualized entities as virtual entities-oriented resource allocation management.

In order to ensure network performance while minimizing power consumption, Caviglione et al. [45] applied a DRL algorithm, named Rainbow DQN, to solve the multi-objective VM placement problem. Their model was based on the percentages of network capacity, CPU, and disk, with full consideration of energy cost, network security, and QoS. Liu et al. [152] applied Q-learning algorithm to distributed management of resources, and their proposed hierarchical network architecture can provide resource allocation and power management of VMs. Experiments showed that when the physical server clusters are set to 30, for 95,000 jobs, the proposed hierarchical framework can reduce the network energy consumption and latency by 16.12 and 16.67% respectively, compared with the DRL-based resource allocation. It can be seen that in addition to the resource allocation for tasks, the resource allocation for virtual entities also greatly affects the operation efficiency and power consumption of the data center. Elprince et al. [75] designed a dynamic resource allocator, which allocated resources through different machine learning techniques (such as REPTree and Linear Regression), and dynamically adjusted the allocated resources through a resource fuzzy tuner. Experiments showed that their solutions can guarantee SLA well and meet differentiated service requirements between various customers. Jobava et al. [125] managed VM resources through flow-aware consolidation. The AL algorithm was used to divide the virtual clusters to reduce the total communication cost, and then simulated annealing algorithm was employed for intelligent allocation of VM clusters. Both phases were traffic aware.

3.4.3 QoS-Oriented Resource Management

Resource management optimization research has improved the overall QoS of the service by optimizing resource allocation, although this is not the primary key objective. The two typical representative research work aiming at QoS are as follows. Wang et al. [275] proposed an on-demand resource scheduling method based on DNN to ensure the QoS of delay-sensitive applications. Wadwadkar et al.

[287] leveraged SVM to perform resource prediction and meet QoS requirements with a performance-aware resource allocation policy. As for the uncertainty of prediction, they introduced the concept of confidence measure to mitigate this problem.

3.4.4 Resource Prediction-Oriented Resource Management

Resource prediction plays an essential role in resource management. Timely and accurate resource forecasting can make the data center achieve more effective resource scheduling, and further improve the overall performance of the data center network. However, although virtualization and other technologies greatly enrich the types of resources and improve service efficiency, it also increases the difficulty of resource prediction. Besides, Aguado et al. [5] implied that the prediction accuracy of traditional unintelligent algorithms cannot be guaranteed on account of diversity of services and bandwidth explosion. Moreover, to cope with the unpredictable resource demand, traditional resource management mechanisms usually over-allocate resources to ensure the availability of resources, which is harmful to the overall resource utilization of data centers. How to deal with the differentiated requirements of various workloads and precisely predict resources still remains a challenge. Yu et al. [295] proposed a deep learning-based flow prediction and resource allocation strategy in optical DCNs, and experimental results demonstrated that their approach achieved a better performance compared with a single-layer NN-based algorithm. Iqbal et al. [121] proposed an adaptive observation window resizing method based on a 4-hidden-layer DNN for resource utilization estimation. The work of Thonglek et al. [251] predicted the required resources for jobs by a two-layer LSTM network, which outperformed the traditional RNN model, with improvements of 10.71 and 47.36% in CPU and memory utilization, respectively.

3.4.5 Resource Utilization-Oriented Resource Management

Resource utilization is regarded as an intuitive and important metric to evaluate a resource management mechanism. This type of resource management schemes typically improve the resource utilization through task scheduling, VM migration and load balancing algorithms. It is worth noting that dynamic change of resource demand in data centers requires the algorithm being able to automatically optimize resource utilization according to the changing network environment. However, the traditional unintelligent solutions are difficult to cope with the high variability of the network environment. Therefore, a few researchers have begun to apply machine learning to solve these problems. Che et al. performed task scheduling based on the actor-critic deep reinforcement learning algorithm to optimize resource utilization and task completion time [48]. Telenyk et al. [246] used the Q-learning algorithm for global resource management, and realized resource optimization and energy saving through virtual machine scheduling and virtual machine aggregation. In addition to improving resource utilization through scheduling and consolidation, Yang et al. [291] focused their research on the optimization of storage resource, that is, how to efficiently store data. They used distributed multi-agent reinforcement learning methods to achieve joint optimization of resources, which effectively improved network throughput and reduced stream transmission time.

3.4.6 Discussion and Insights

The current resource management system in today's data centers is complex and multifaceted. Along with the expansion of service scenarios, the resource scheduling among various virtualized entities is getting more complicated. Increasingly, researchers adopt deep learning or deep reinforcement learning aiming to achieve a more intelligent resource management. We list the details of each intelligent solution in Table 3.5. Then, we evaluate these solutions with respect to each dimension of REBEL-3S, as shown in Table 3.6. It reveals that more than half of the solutions take the energy efficiency into account in resource management [45, 163, 246], and most of them consider the impact of network stability. Here, we summarize several key concerns, as below, which deserve to be further studied and addressed.

- **Stability and scalability of models.** Taking reinforcement learning as an example, primary decisions may have relatively poor consequences due to a lack of domain knowledge or good heuristic strategies [247]. When the agent performs tentative actions, it may fall into local optimal solutions if not appropriately trained. Besides, reinforcement learning may lack good scalability in large DCNs.
- **Adaptability to Multi-objective and multi-task.** Whether it is a traditional resource allocation scheme (such as priority-based VM allocation [236], heuristic-based resource allocation [36]), or an intelligent resource allocation scheme, their performance is usually evaluated in a specific single scenario. Whereas, one qualified intelligent solution should fully consider the richness of scenarios and requirements, and be able to adapt to multi-scenario and multi-task network environment.
- **Security of resource allocation.** The flexibility of virtualized resources can make vulnerability or fault propagation faster, and fault recovery and fault source tracing more difficult. Padhy et al. [187] disclosed that vulnerabilities were found in VMware's shared folder mechanism, which could allow the users of guest systems to read and write to any part of the host file system, including system folders and other security-sensitive files.
- **Perspective of resource lifecycle.** The allocation, utilization, and recycling of resources occur frequently. Current intelligent solutions focus more on the

Table 3.5 Research progress of data center network intelligence: resource management

Ref	Category[a]	ML category & model adopted	Features	Data source	Feature selection	Additional constraints	Estimation function	Experimental comparison subjects
Wang et al. [275]	Q	DL, DNN	Monitor the performance of applications in real time, and adjust the resource allocation policy of the corresponding app if performance is impaired to ensure QoS	A web search engine from TailBench benchmark	Internal and external performance counter data	QoS	CPU Quota, etc.	✓|✓|✗
Che et al. [48]	RU	DRL, Actor-critic	Task Scheduling, with the goal of optimizing resource utilization and task completion time	Alibaba Cluster Trace Program [10]	Task and VM Properties	None	Average task latency time, average task priority, task congestion level	✓|✓|✓
Liu et al. [152]	V	DRL, Q-learning	Using a hierarchical network architecture for VM resource allocation and power management	Real-world data center workload traces from Google cluster-usage traces [94]	Task arrival time, task duration, task resource requests (CPU, memory, disk requirements)	Minimize energy consumption and maintain reasonable performance	Average delay, etc.	✓|✓|✓

Reference	Type	Algorithm	Objective	Dataset	Features	Privacy	Metrics	
Tesauro et al. [247]	T	DRL, DNN/Sarsa	Allocate server resources to applications	Simulated data	Application and server properties	None	SLA Total Revenue, etc.	✓|✓|✓|✗
Liu et al. [155, 157]	T	DRL, DQN	Reduce service latency with data placement policies	MSR Cambridge Traces [183]	End-to-end node information, read/write latency, subsequent analysis latency, etc.	Network Latency	Average data read latency, etc.	✓|✓|✓|✓
Yang et al. [291]	RU	RL, Q-learning	The storage mode of resources was studied to realize the joint optimization of data storage and traffic management	Simulated data	Transfer rate, remaining server capacity, etc.	None	Transmission completion time, total number of dropped packets, etc.	✓|✓|✓|✗
Iqbal et al. [121]	RP	DL, DNN	Predicting resource utilization and proposing a resource estimation model	Three publicly available datasets [10, 100, 101]	VM CPU, memory, network and disk utilization.	None	MSE	✓|✓|✓|✓
Elprince et al. [75]	V	SL/UL, REPTree/Linear Regression, etc.	Allocate resources to VCs based on ML	Real-world traces of Los Alamos National Lab	Detailed information about resource requests and usage, such as memory and CPU time	Guaranteed SLA	RMSE, RAE, RRSE, etc.	✓|✗|✓|✓

(continued)

Table 3.5 (continued)

Ref	Category[a]	ML category & model adopted	Features	Data source	Feature selection	Additional constraints	Estimation function	Experimental comparison subjects
Caviglione et al. [45]	V	DRL, DQN	Used DRL to solve multi-target VM placement problem	Self-built data center	Percentage of CPU, disk and network requested by VM, etc.	Energy overhead, QoS, network security	Customized	✓|✓|✗
Yi et al. [294]	T	DRL, DQN	Allocate server resources to applications	Simulated data	Processor utilization, temperature, power consumption, number of spare cores	Minimize power consumption and maintain reasonable performance	Processor temperature and power consumption, etc.	✓|✓|✓
Li et al. [163]	RU	SL, M5P[104]	Modeling training on resources / QoS / Workload based on historical data to help develop better scheduling algorithms to better balance throughput, QoS and energy efficiency	Realistic workloads and environments	Average computation time per request and average number of bytes exchanged per request, etc.	Energy overhead, QoS	Service level agreement fulfillment, etc.	✓|✓|✓

Thonglek et al. [251]	RP	DL, LSTM	Predicting job requirements	Google dataset [178]	Requested and used CPU, memory resources	None	CPU and memory utilization	✓|✗|✓
Xu et al. [285]	RU	RL, Unspecified	Blockchain-based resource provisioning	Google dataset [150]	Request resources needed for migration, such as CPU cores, RAM, Disk, etc.	Energy overhead	Energy cost, etc.	✓|✓|✗
Zerwas et al. [303]	RP	SL/UL, CNN/LR, etc.	An ML framework for predicting virtual cluster acceptance rates	Simulated data	Link capacity and free resources, etc.	None	Calculation of unit utilization, etc.	✓|✓|✓
Chen et al. [55]	T	RL, Actor-critic	Allocate resources to jobs (model-free RL), minimizing task latency	Google dataset	Assignment completion time, waiting time, etc.	Network Latency	Normalized latency, etc.	✓|✓|✓
Yu et al. [295]	RP	DL, DNN	A deep learning-based resource allocation algorithm	Self-built data center	Resource usage status and prioritization of existing flow	None	Path blocking probability, etc.	✓|✓|✓

(continued)

Table 3.5 (continued)

Ref	Category[a]	ML category & model adopted	Features	Data source	Feature selection	Additional constraints	Estimation function	Experimental comparison subjects
Yadwadkar et al. [287]	Q	SL, SVM	Multi-dimensional solutions	Multiple Hadoop deployments (including deployments for Facebook and Cloudera customers)	CPU utilization, disk utilization and other information	QoS	Accuracy, etc.	✔|✔|✘
Telenyk et al. [246]	RU	RL, Q-learning	Operating with Global Manager for VM scheduling, VM consolidation, etc. for resource optimization, energy saving	Bitbrains [277]	CPU, memory and network bandwidth	Energy overhead, QoS	Number of SLA violations, etc.	✔|✔|✘
Jobava et al. [125]	V	RL, AL	VM resource management through flow-aware consolidation	Simulated data	Cost matrix	None	Total cost of communication, etc.	✔|✘|✘

Marahatta et al. [170]	T	RL, DNN	Allocate resources for tasks (predict tasks based on ML, and divide the tasks into error-prone and non-error-prone tasks, and implement different allocation strategies)	Eular dataset and Internet dataset	The requested resources of the tasks, the actual allocation of resources and whether failure occurred	Energy overhead, QoS	Task failure rate, etc.	✓ | ✓ | ✗
Wang et al. [265, 268]	V	FTR, BI	The first to apply BI paradigm to solve the VNE problem	Simulated data	CPU capacity and bandwidth capacity between nodes, etc.	Minimize embedding costs and improve economic revenue	Customized	✓ | ✓ | ✗
Rayan et al. [208]	RP	SL, SVM, etc.	Workload prediction based on ML, based on which the required resources can be predicted	Simulated data	Number of active physical machines and power consumption data	Energy overhead	RMSE, running time	✓ | ✓ | ✓

[a]For convenience, we use "T" for Task-oriented Resource Management, "V" for Virtual Entities-oriented Resource Management, "Q" for QoS-oriented Resource Management, "RP" for Resource Prediction-oriented Resource Management, and "RU" for Resource Utilization-oriented Resource Management

Table 3.6 Assessment of resource management schemes based on REBEL-3S

Ref	Reliability	Energy efficiency	Bandwidth utilization	Latency	Security	Stability	Scalability
Wang et al. [275]	YES	NO	YES	YES	NO	YES	NO
Che et al. [48]	NO	NO	NO	YES	NO	YES	NO
Liu et al. [152]	NO	YES	YES	YES	NO	YES	NO
Tesauro et al. [247]	NO	NO	YES	YES	NO	YES	NO
Liu et al. [155, 157]	YES	YES	YES	YES	NO	YES	YES
Yang et al. [291]	NO	NO	YES	YES	NO	YES	YES
Iqbal et al. [121]	NO	NO	YES	YES	NO	NO	NO
Elprince et al. [75]	NO	NO	YES	YES	NO	NO	YES
Caviglione et al. [45]	NO	YES	YES	YES	YES	YES	NO
Yi et al. [294]	NO	YES	YES	YES	NO	YES	YES
Li et al. [163]	NO	YES	YES	NO	NO	NO	NO
Thonglek et al. [251]	NO	NO	YES	NO	NO	YES	NO
Xu et al. [285]	NO	YES	YES	NO	NO	YES	NO
Zerwas et al. [303]	NO	YES	YES	YES	NO	NO	NO
Chen et al. [55]	NO	NO	YES	YES	NO	YES	NO
Yu et al. [295]	NO	NO	YES	YES	NO	NO	NO
Yadwadkar et al. [287]	YES	YES	YES	YES	NO	YES	YES
Telenyk et al. [246]	NO	YES	YES	YES	NO	YES	NO
Jobava et al. [125]	YES	YES	YES	YES	NO	YES	YES
Marahatta et al. [170]	YES	YES	YES	NO	NO	YES	NO
Wang et al. [265, 268]	YES	YES	YES	NO	NO	NO	YES
Rayan et al. [208]	NO	YES	YES	NO	NO	NO	NO

prediction of resource allocation and maximization of benefits in the process of resource allocation, but lack related studies on resource collection and recycling.

3.5 Energy Management

With the rapid rise of cloud-based service needs, both the computing load and traffic load of data centers increase exponentially. In this case, any kind of unreasonable resource planning (e.g. unbalanced load, over-provisioned resources, non-energy-aware routing, etc.) will result in serious energy waste. Therefore, energy efficiency optimization has become an urgent and crucial issue for sustainable data centers, and many researchers have conducted considerable research in power conservation of data centers.

Traditional energy management solutions can be divided into model-based and state-based solutions [161]. Model-based solutions require designing corresponding mathematical models for each hardware module, such as the air supply temperature of the air conditioning mainframe, the CPU power consumption of the server, and the cooling ducts' inlet and outlet temperatures. Most model-based solutions are based on computational fluid dynamics (CFD) for simulation, which can find the optimal approximate solution, however, it requires high computing power and memory capacity, and is very time-consuming. Additionally, the modeling parameters also need to be adaptively updated in line with the network conditions. Toulouse et al. [254] proposed the potential-flow-based compact model as an alternative to CFD. However, it only retained the most basic physical mechanisms in the flow process and could not meet the need for real-time control. Besides, these representative schemes were based on a relatively stable premise and may not perfectly fit real-world scenarios of data center networks. State-based solutions are more suitable for real-time control, which take the historical state information of the device as input and use machine learning algorithms to control power consumption. Parolini et al. [192] introduced a state-space model to predict temperature changes dynamically. Based on the air conditioning supply temperature and rack power, Tang et al. [243] constructed an abstract heat flow model to calculate the temperature distribution of the rack.

Positively, machine learning provides a promising way to deal with these challenges faced by the traditional models. The rest of this section will sequentially introduce the current research on ML-based energy management at the server level, network level, and data center level, and finally give our insights.

3.5.1 Server Level

The server level energy optimization concentrates on the aggregation of services leveraging VM migration, VM mapping, and resource prediction/scheduling, which pays little attention to network-level performance. Marahatta et al. [169] used deep neural networks to classify tasks according to the probability of task failure, and proposed an energy-aware scheduling method based on failure prediction model to achieve lower energy consumption with high fault tolerance. The data-driven NN-based framework designed by Uv et al. could perform power prediction based on server performance data [260]. Yang et al. [289] fully considered the uncertainty of the network and the diversity of task characteristics, and transformed the economic energy scheduling problem into a Markov decision process and based on which an energy-aware resource scheduling scheme was proposed aiming to reduce the energy consumption in data centers.

3.5.2 Network Level

The key principle of the network level energy management is to calculate a subset of the network to carry the traffic through flow aggregation, flow scheduling, energy-aware routing, etc., and shut down the unused network devices so as to maximize the power savings. Wang et al. [272] were the first group to apply Blocking Island (BI) paradigm, which is an ML-based resource abstraction model, to the decision-making process of energy-saving strategies from the perspective of resource allocation and routing. The experimental results revealed that their approach greatly improved the computation efficiency and achieved at most 50% power savings while guaranteeing the network reliability. Sun et al. [240] proposed a flow scheduling method that employed deep reinforcement learning to reduce network energy consumption while ensuring FCT. They argued that the optimal flow scheduling scheme in DCNs should consider the temporal fluctuations and spatial distribution imbalances of flow when it dynamically merges traffic to fewer active links and devices, so as to improve the power efficiency. Inspired by these research work, we hold that the goal of energy conservation solutions can be achieved from two dimensions: local dimension and overall dimension. The local dimension needs to ensure a fine-grained flow distribution adjustment policy on each path, while the overall dimension needs to implement a dynamic traffic adjustment policy and an FCT assurance policy based on real-time flow fluctuations.

3.5.3 Data Center Level

This level is more complex and systemic. It is dedicated to the energy savings of the whole data center, including not only network devices and computing equipment, but also related supporting systems, such as air conditioning cooling systems, heat dissipation systems, electrical power supply systems, and so forth. Athavale et al. [21] used cooling system data, such as the computer room air conditioner (CRAC) blower speed and CRAC return temperature setpoint, to correlate with the DNN model's flow and temperature distributions. They claimed that the solution was able to anticipate temperature and tile flow rate with a normal blunder of <0.6 °C and 0.7%, respectively. Actually, their DNN model based on a transient scenario may had lower interpretive prediction errors, but higher extrapolative prediction errors. Grishina et al. [96] analyzed the Information Technology (IT) room's thermal characteristics by clustering them with the K-means algorithm, effectively circumventing the drawbacks of the traditional temperature measurement solutions, for example, the difficulty of locating specific nodes causing overheating. Apart from the assessment of characteristics, researchers also paid attention to the assessment of risk. Sasakura et al. [218] used the gradient boosting decision tree (GBDT) and a state space model to predict the temperature and executed different energy-saving strategies by evaluating the risks associated with the temperature. Furthermore, to solve the time-consuming problem of the traditional CFD parameter identification process, Fang et al. [79] performed the relationship mapping between traffic patterns and model parameters through DNN. Their scheme primarily took the flow rate, rack power distribution, and air supply temperature of the server room air conditioner into account for the rapid temperature assessment. Ran et al. [206] focused on the joint optimization of job scheduling and cooling control in data centers to further improve energy efficiency. The proposed dual-scale control mechanism provided a good tradeoff between QoS and energy conservation.

Table 3.7   Assessment of energy management schemes based on REBEL-3S

Ref	Reliability	Energy efficiency	Bandwidth utilization	Latency	Security	Stability	Scalability
Grishina et al. [96]	NO	YES	YES	NO	NO	YES	YES
Fang et al. [79]	NO	YES	YES	NO	NO	YES	YES
Liu et al. [161]	NO	YES	YES	NO	NO	YES	YES
Sasakura et al. [218]	NO	YES	YES	NO	NO	YES	NO
Wang et al. [272]	YES	YES	YES	NO	NO	YES	YES
Athavale et al. [21]	YES	YES	YES	NO	NO	YES	NO
Yang et al. [289]	NO	YES	YES	YES	NO	YES	NO
Ran et al. [206]	NO	YES	YES	YES	NO	YES	NO
Kuwahara et al. [134]	NO	YES	YES	NO	NO	YES	YES
Uv et al. [260]	YES	YES	YES	YES	NO	YES	NO
Marahatta et al. [169]	YES	YES	YES	NO	NO	YES	NO
Merizig et al. [174]	NO	YES	YES	NO	NO	NO	NO
Haghshenas et al. [102]	YES	YES	YES	YES	NO	YES	YES
Jobava et al. [125]	YES	YES	YES	YES	NO	YES	YES
Sun et al. [239, 240]	NO	YES	YES	YES	NO	YES	YES
Ilager et al. [115]	NO	YES	YES	YES	NO	YES	NO
Li et al. [144]	NO	YES	YES	YES	NO	YES	NO
Shoukourian et al. [233]	YES	YES	YES	YES	NO	YES	NO

Nevertheless, this scheme was specially designed for computing intensive tasks, and had limitations in application scenarios. Besides QoS, Kuwahara et al. [134] also regarded the seasonal factor as a constraint in their scheme, where they held that seasonal climate change was critical to the temperature impact of data centers. Hereby, they collected power consumption data for about one year and concluded that the power consumption of PACs varied greatly in summer and winter. Their solution was based on the CFD model to build the power consumption model of the device in the early stage, and then used deep learning to predict power consumption. As the authors claimed, compared with repeatedly building CFD based power consumption model on a single device to predict power consumption, the process efficiency and accuracy were improved by 900 and 8%, respectively. However, it did not work well for on-demand task assignments. Thus, they conducted further research [135] to improve this situation (Table 3.7).

3.5.4   Discussion and Insights

We presented some of the most representative existing research works under different categories, and briefly elaborated the key characteristics of each scheme, as summarized in Table 3.8. Moreover, as revealed in Table 3.7, all these schemes were further evaluated and compared from various dimensions according to the REBEL-3S criteria. As for the experimental comparisons and evaluations of these schemes,

Table 3.8 Research progress of data center network intelligence: energy management

Ref	Category[a]	ML category and model adopted	Features	Data source	Feature selection	Additional constraints	Estimation function	Experimental comparison subjects
Grishina et al. [96]	D	UL, K-means	Continuous cluster thermal characterization by clustering	Monitoring data of the CRESCO6 cluster in the premises of ENEA-Portici Research Center (R.C.)	Exhaust temperature, CPU temperature readings, etc.	None	Clustering size and intersection ratio of node labels, etc.	✓|✗|✗
Fang et al. [79]	D	DL, DNN	Heat assessment framework based on thermal model and traffic pattern	A small data center at Tongji University, Shanghai, China	Node entrance and exit temperatures, airflow rates, etc.	None	MAE, MAPE, RMSE	✓|✓|✗
Liu et al. [161]	N	DL, LSTM/CNN	Different coarse-grained power consumption prediction based on uniform character set encoding strategy	World Cup 98 (WC98) dataset [20]	Historical power consumption, historical RUI requests, etc.	None	MSE	✓|✗|✓
Sasakura et al. [218]	D	SL, GBDT	Power consumption prediction and regulation based on DT model	Simulated data	Power consumed by switch fixed overhead, port power consumption, etc.	None	Customized	✓|✓|✗

	N/D/S	Method	Description	Data source	Features	Constraint	RMSE, etc.			
Wang et al. [272]	N	FTR, BI	Applied the Blocking Island Paradigm for resource allocation into DCNs to achieve power conservation	Self-built data center	Rack CRAC and intake air temperature sensor and other related data	None	RMSE, etc.	✓	✓	✓
Athavale et al. [21]	D	DL, DNN	Data center temperature and flow forecasting via DNN	Simulated data	Server room air conditioning blower speed, IT load distribution of racks, etc.	None	MSE	✓	✓	✗
Yang et al. [289]	S	DRL, DDPG	Markov decision based network energy consumption optimization	Real-world data sets [66, 73, 95]	Workload, queue length, battery capacity, etc.	None	Customized	✓	✓	✗
Ran et al. [206]	D	DRL, DQN	Adopt dual scale control mechanism	Simulated data	Number of CPUs available to the server, utilization, power consumption, outlet temperature, etc.	QoS	Customized	✓	✓	✓
Kuwahara et al. [134]	D	DL, DNN	NN training based on CFD	Simulated data	Server fan speed, etc.	Seasonality	Total power consumption, etc.	✓	✓	✗

(continued)

Table 3.8 (continued)

Ref	Category[a]	ML category and model adopted	Features	Data source	Feature selection	Additional constraints	Estimation function	Experimental comparison subjects
Uv et al. [260]	S	RL, Q-learning	Power prediction based on server performance data	IBM System Z Partition	Server load, temperature, etc.	QoS	MSE	✓|✗|✗
Marahatta et al. [169]	S	DL, DNN	Resource provisioning by task classification to reduce total energy consumption	Real-world data sets	CPU utilization, etc.	None	Total energy consumption, etc.	✓|✓|✗
Merizig et al. [174]	S	UL/DL, SVM/DNN	A comparison of several ML methods was presented	Unknown data center	Unspecified	None	Energy consumption	✓|✓|✓
Haghshenas et al. [102]	S	RL, Q-learning	Energy management through VM scheduling	Unknown data center	Data metrics of physical hosts, VMs, and data center hardware equipment	QoS	Number of SLA violations, number of VM migrations, etc.	✓|✓|✗

Reference	Level	Method	Description	Dataset	Cost matrix		Evaluation	S	N	D
Jobava et al. [125]	N	RL, LA	VM resource management through flow-aware consolidation	Simulated data	Cost matrix	None	Total cost of communication, etc.	✓	✗	✗
Sun et al. [239, 240]	N	DRL, DDPG	The DRL algorithm was used to improve power efficiency and ensure FCT	Wikipedia trace files [259]	Number of switch ports, port power, incoming and outgoing flow rates, etc.	QoS	FCT, etc.	✓	✓	✗
Ilager et al. [115]	S	SL, DT	A gradient boosting machine learning model for temperature prediction was proposed	Datasets for Bitbrain [228].	Host-level resource data and sensor readings	Rigid thermal	SLA violation number, temperature distribution, etc.	✓	✓	✗
Li et al. [144]	D	DRL, DDPG	An end-to-end cooling airtime algorithm was proposed	Singapore National Supercomputing Center (NSCC).	Entrance and exit temperatures, etc.	None	Energy saving, etc.	✓	✓	✓
Shoukourian et al. [233]	D	DL, DNN	Energy management based on data center cooling system data	PowerDAM [232]	The amount of cold aggregation generated by the cooling loop and the accumulated power consumed, etc.	None	Accuracy etc.	✓	✓	✗

[a] For convenience, we use "S" for Server Level, "N" for Network Level, and "D" for Data Center Level

most of them were evaluated and compared in terms of the prediction accuracy of temperature and power consumption in addition to the effect of energy savings. Hereby we summarize several key concerns and open issues, as follows.

- **Data acquisition.** Since most intelligent solutions are data-driven, the impact of data acquisition process on the performance and security of server and network cannot be ignored. To mitigate this issue, most of the researchers [21, 125, 134, 169, 206, 239, 289] simply used the offline ready-made or simulated data sets, which cannot prove the performance in the real world. Therefore, it is quite necessary to explore an efficient and low cost data acquisition mechanism, and provide an effective quantitative method to quantify the impact of data acquisition on related facilities.

- **Data analysis.** The survey statistics indicate that NN is the most commonly used model in the existing work. However, there is a lack of relevant research on data interdependencies and interactions, which are crucial in NN feature engineering.

- **Applicability.** Hyperparameter tuning is not only an essential but inevitable procedure in the process of NN training. However, for the dynamics problem with different configurations in a complex data center system, it is found that it lacks certain applicability. This is an interesting and significant research topic, which deserves more attention and research findings.

3.6   Routing Optimization

In DCNs, routing optimization is one of the most important research areas and has aroused some discussions in both academia and industry. With the advantage of SDN, routing optimization can get a global view of the network and deploy strategies conveniently, but the existing traditional SDN-based methods cannot sensitively adapt the real-time traffic changes in data center networks [16, 99, 149, 266, 267, 282]. For instance, if the routing policies cannot be timely adjusted according to the dynamic network conditions, the imbalance of network flows may cause uneven load distribution among network nodes, where some nodes are highly loaded or even overloaded while some other nodes are underutilized or even idle, resulting in the waste of resources and the degradation of QoS.

The rise of ML techniques has brought new thinking to this field. Chen et al. [52, 53] pointed out that ML-based routing schemes can efficiently solve path optimization problems in complex dynamic network environments, while traditional unintelligent routing schemes are difficult in achieving similar results under the same conditions. However, our investigation shows that so far there has been relatively little research in this field, and most of the existing ML-based routing solutions in data centers are centralized schemes based on SDN. In this book, we divide the small amount of existing ML-based research work into intra-DC and inter-DC routing optimization.

3.6.1   Intra-DC Routing Optimization

About 80% of researchers paid attention to the routing optimization within a data center. Bolodurina et al. proposed [40] a routing optimization strategy taking the SLA into consideration. They pre-classified communication channels based on unsupervised learning and deep learning to obtain a detailed network feature set and then performed clustering according to SLA requirements. Finally, the network feature set and clustering data were used as input for NN training to obtain a suitable routing strategy. Likewise, Fu et al. [86] also adopted a pre-classification strategy before using deep Q-learning (DQL) to train different NNs for elephant flows and mice flows, respectively. They computed optimal routing paths for different types of flows with the help of SDN, ensuring low latency for mice flows and high throughput for elephant flows. Nevertheless, this solution suffers high computation overhead and requires a relatively long time in path calculation. Yu et al. [296] used the DDPG algorithm for routing decisions and improved the network performance providing stable and high-quality routing services. It achieved good convergence and effectiveness while guaranteeing QoS (delay minimization and throughput maximization). Yao et al. [292] designed a DQN-based energy-aware routing algorithm to find energy-efficient data forwarding paths and control paths for switches in data centers.

3.6.2   Inter-DC Routing Optimization

Routing optimization among data centers has also received much attention. Hong et al. [113] used NN to predict flow load blocking probability and proposed an efficient routing and a resource allocation policy among data centers. Comparatively, Francois et al. [84] proposed a logically centralized cognitive routing engine (CRE) based on reinforcement learning and deep learning to meet SLAs through a cognitive routing engine. The CRE can work well even in highly chaotic environments, where it can leverage RNN with RL to find the efficient overlay paths with minimal monitoring overhead among geographically dispersed data centers. Panda and Satyasen [188] combined the adaptive ant colony optimization algorithm with the neural network to solve the power consumption and routing optimization problems in the elastic optical data center network (Table 3.9).

3.6.3   Discussion and Insights

We list details of each intelligent solution in Table 3.10, and evaluate each solution from various dimensions according to REBEL-3S in Table 3.9. Besides, we summarize a general ML-assisted routing framework based on the investigation

Table 3.9 Assessment of routing optimization schemes based on REBEL-3S

Ref	Reliability	Energy efficiency	Bandwidth utilization	Latency	Security	Stability	Scalability
Bolodurina et al. [40]	NO	NO	YES	YES	NO	YES	NO
Fu et al. [86]	NO	NO	YES	YES	NO	YES	NO
Yao et al. [292]	NO	YES	YES	YES	NO	YES	NO
Yu et al. [296]	NO	NO	YES	YES	NO	YES	NO
Panda [188]	NO	YES	YES	YES	NO	YES	NO
Zhou et al. [311]	NO	NO	YES	YES	NO	YES	NO
Liu et al. [148, 162]	NO	NO	YES	YES	NO	YES	YES
Liu et al. [158]	NO	NO	YES	NO	NO	YES	YES
Hong et al. [113]	NO	NO	YES	NO	NO	YES	NO
Francois et al. [84]	NO	NO	NO	YES	YES	YES	YES

results (as shown in Fig. 3.2). The intelligent routing decision module is mainly implemented in the centralized controller. The control plane collects real-time network operation data and performance data through the OpenFlow protocol, and then take advantage of ML to train and calculate the efficient routing policies. Finally, these network policies are distributed to the network devices for execution through the SDN southbound interface. Based on the limited research work in this area, we summarize the following several concerns as well as some potential research opportunities.

- **Multi-objective optimization.** Considering the diversity of traffic types with different priorities, how to provide differentiated routing optimization policies is deemed as a challenging task. What's more, the decision making of routing policies may also depend on the preliminary results of some other optimization models, such as traffic identification and classification schemes, where it will involve collaborative optimizations of multiple learning models for multiple tasks, which further increases the difficulty of optimization. In addition, the optimization objectives are often diverse as well, and even need to be satisfied at the same time (such as high throughput, low latency, high reliability, load balancing, high link utilization, fault tolerance/burst tolerance, and even high energy efficiency, etc.), which poses a great challenge to the optimality of the solutions (global optimal or local optimal) and the computational complexity of algorithms.
- **ML model selection.** Choosing an appropriate and effective ML model is the first and most important step towards intelligent routing optimization. However, in view of the particularity of the data center scenario, in most cases, the existing ML models can not be applied directly to the network routing optimization. For instance, the original Q-learning algorithm, which can only deal with discrete action problems, cannot handle the continuous dynamic changes in the network, and the complex and diverse network states may lead to excessive storage space for Q-table. In consideration of the fact that the research work on ML-based routing optimization is still relatively little, the types of ML models used in the existing schemes are relatively few. As of now the investigation shows that

Table 3.10 Research progress of data center network intelligence: routing optimization

Ref	Category[a]	ML category and model adopted	Features	Data source	Feature selection	Additional constraints	Estimation function	Experimental comparison subjects
Bolodurina et al. [40]	A	UL/DL, DNN	Optimized routing based on network characteristics and QoS requirements	Simulated data	Channel delay, jitter, packet loss rate, QoS rules used, etc.	QoS	Resource capacity and network response time	✓|✓|✗
Fu et al. [86]	A	DRL, DQN/CNN	Efficient routing strategies for elephant and mice flows, respectively	Simulated data	Source switch, destination switch and bandwidth requirements for streams, etc.	Low latency and low packet loss rate for mice streams and high throughput and low packet loss rate for elephant streams	Customized	✓|✓|✗
Yao et al. [292]	A	DRL, DQN	An energy-efficient load balancing strategy	Simulated data	Source node, destination node and bandwidth requirements of the flow, etc.	Energy consumption	Customized	✓|✓|✗

(continued)

Table 3.10 (continued)

Ref	Category[a]	ML category and model adopted	Features	Data source	Feature selection	Additional constraints	Estimation function	Experimental comparison subjects
Yu et al. [296]	A	DRL, DDPG	DDPG-based routing policy with excellent convergence and effectiveness	Simulated data	Latency, throughput, etc.	QoS	Network latency, throughput	✓|✓|✗
Panda [188]	B	DL, DNN	Solving power consumption and routing optimization problems of elastic optical networks based on NNs and adaptive ant colony optimization algorithms	Simulated data	Input weight matrix of encoder, etc.	Energy consumption	Spectrum utilization, etc.	✓|✓|✗
Zhou et al. [311]	A	DL, RNN	A residual flow compression mechanism was introduced to minimize the completion time of data-intensive applications	Simulated data	Coflow width, coflow size and arrival time, etc.	Network bandwidth	FCT, etc.	✓|✓|✗

| Liu et al. [148, 162] | A | DRL, DQN/DDPG | The DRL-R (Deep Reinforcement Learning-based Routing) algorithm was proposed to bridge multiple resources (node cache, link bandwidth) by quantifying the contribution of multiple resources (node cache, link bandwidth) to reduce latency | Simulated data | Node cache, link bandwidth | Maximize overall network throughput while meeting QoS | FCT, etc. | ✔ ✔ ✗ |
| Liu et al. [158] | A | DRL, DDPG/CNN | A hybrid flow scheduling scheme based on deep reinforcement learning was proposed | Simulated data | Paths and flows information | Maximizing deadline satisfaction rate for mice flow and minimizing FCT for elephant flow | Deadline meet rate [153] | ✔ ✔ ✗ |

(continued)

Table 3.10 (continued)

Ref	Category[a]	ML category and model adopted	Features	Data source	Feature selection	Additional constraints	Estimation function	Experimental comparison subjects
Hong et al. [113]	B	DL, DNN	An efficient routing and resource allocation strategy between data centers.	Simulated data	Number of racks, average capacity required for connection requests on racks, etc.	None	CDF	✔|✘|✘
Francois et al. [84]	B	DL/RL, RNN	A logical centralized cognitive routing engine was developed based on stochastic NNs with reinforcement learning	5 geographically dispersed data centers	Latency	QoS	RTT etc.	✔|✔|✘

[a] For convenience, we use "A" for Intra-DC Routing Optimization and "B" for Inter-DC Routing Optimization

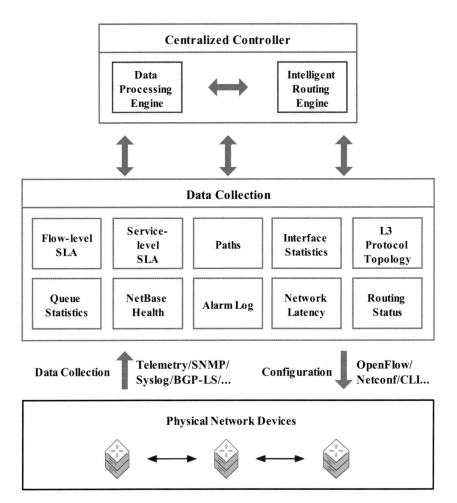

Fig. 3.2 A general ML-assisted routing framework

deep reinforcement learning is still the mainstream paradigm for solving routing optimization problems. Thereby, it is deemed that more effective ML models need to be explored and validated, which is of great necessity and importance.

3.7 Congestion Control

The complexity and diversity of service scenarios and finer granularity of flow demands have made congestion control more complicated in data centers. For instance, some applications require high micro-burst tolerance [225, 226], while some applications demand low latency [176] or high throughput [88]. Besides, the diverse applications and computing frameworks with different characteristics in data centers further produce a variety of traffic patterns, such as one-to-one, one-to-many, many-to-one, many-to-many, and all-to-all traffic patterns. However, the traditional TCP-based solutions can hardly meet all the requirements of these different traffic patterns at the same time [70, 83], which often result in queueing delay, jitter incast, throughput collapse, increased flow completion time, and packet loss [58, 164, 167].

Admittedly, congestion control (CC) is the core of the TCP algorithm, which determines the data transmission efficiency. Although the research on CC has spanned more than three decades, the vast majority of CC solutions in data center network scenarios have followed a conservative strategy. It starts transmission at a slow sending rate and then uses certain strategies (e.g., AIMD) to adjust the sending rate during subsequent transmissions, which is normally agnostic to the flow deadline and network congestion and cannot well cope with the micro-burst scenario as well [57, 127]. For instance, when multiple synchronous servers send data to a single receiver simultaneously, the shallow-buffered switches at the last hop are prone to be overwhelmed by the bursty traffic resulting in increased queueing delay or even packet loss, which is known as the TCP incast problem [237]. Explicit Congestion Notification (ECN) is the most common congestion handling mechanism, and MQ-ECN [25] is the first protocol to enable multi-queue scenarios in data centers, guaranteeing queue independence to ensure no loss of network latency and throughput. Nevertheless, the traffic is inherently bursty in DCNs, and MQ-ECN is only applicable to round-based scheduling mechanisms.

Recently, ML has attracted researchers' interest, and some ML-based CC algorithms have been proposed. From the perspective of the decision-making mode of CC policies, we divide the existing ML-based solutions into centralized and distributed congestion control.

3.7.1 Centralized Congestion Control

The centralized scheme detects, avoids, and mitigates network congestion through unified scheduling and centralized management of decentralized network resources. The centralized allocation of network resources can maximize overall network resource utilization, but it may also pose some problems. The transmission of relevant network logs will take up additional bandwidth and memory, and the relatively long response time of network policy decisions will have an adverse impact on the latency sensitive applications. What's more, in order to achieve efficient congestion control, many solutions require to customize hardware, which has lost the generality and practicality. More importantly, the centralized congestion control scheme typically suffers from the scalability issue, where the centralized controller usually becomes the bottleneck, which is difficult to adapt to large-scale data center network.

Jin et al. [124] designed two congestion control methods based on the improved Q-learning algorithm and Sarsa algorithm, respectively, with the help of a centralized SDN controller. They believed that the congestion control algorithm should consider the temporal and spatial characteristics of the flows and focused on the path of the current flow when selecting the action. However, the improved algorithms were not competent to adapt to complex data center networks [123]. It is necessary to further optimize the design of the reward function (for example, adding delay, power consumption, fairness, reliability, and other factors into the evaluation dimension), and to test under more complex network environment. Ruffy et al. [213] proposed Iroko, a scalable and modular simulation simulator based on deep reinforcement learning that supported various congestion control algorithms. However, Iroko is still not suitable for large-scale networks, and it cannot adapt to undefined network topologies, where the topology needs to be manually specified, which is not practical in real world scenarios.

3.7.2 Distributed Congestion Control

Compared with centralized schemes, distributed schemes decentralize the decision-making authorities and focus more on end-to-end congestion control, concentrating the collaborative algorithm design on distributed network devices and hosts. Majidi et al. used deep learning to improve the processing capability of switches, and to separate elephant flows and mice flows through dual-coupled queues to meet different FCTs [168]. In addition, the ECN threshold of each queue was dynamically tuned to absorb micro-bursts. Nie et al. [185] proposed a TCP-RL system based on reinforcement learning, which used different learning processes and congestion control strategies for long and short flows to reduce RTT and maximize the overall network throughput. The scheme has been deployed to one of the worldwide top search engines for many years. Whereas, this approach was only evaluated and compared in a static network condition rather than a dynamic DCN (Table 3.11).

3.7.3 Discussion and Insights

A comparative analysis of some typical representative research works is summarized in Table 3.12, and the Table 3.11 shows the evaluation results of each solution according to REBEL-3S. In the light of investigation and evaluation results, we summarize several key problems that could hinder the improvement and implementation of congestion control schemes, as below.

- **Latency.** According to reports provided by cloud service providers [202], slight service delays can cause a dramatic drop in user experience, resulting in significant revenue loss. Therefore, ML-based solutions should maximize the

Table 3.11 Assessment of congestion control schemes based on REBEL-3S

Ref	Reliability	Energy efficiency	Bandwidth utilization	Latency	Security	Stability	Scalability
Jin et al. [124]	NO	NO	YES	NO	NO	YES	YES
Liao et al. [145, 146]	NO	NO	YES	YES	NO	YES	YES
Liu et al. [155, 157]	YES	YES	YES	YES	NO	YES	YES
Majidi et al. [168]	NO	YES	YES	YES	NO	YES	YES
Nie [185]	NO	NO	YES	YES	NO	YES	YES
Ruffy et al. [213]	NO	NO	YES	YES	NO	YES	YES
Thiruvenkatam et al. [250]	NO	NO	YES	YES	NO	YES	YES
Sun et al. [239, 240]	NO	YES	YES	YES	NO	YES	YES
Rastegarfar et al. [207]	YES	YES	YES	YES	NO	YES	YES
Xiao et al. [283]	YES	NO	YES	YES	NO	YES	YES

user experience by speeding up convergence speed (e.g., using asynchronous components, distributed solutions) and shortening latency as much as possible.

- **Stability of ML algorithms.** The flexibility of ML algorithms is a double-edged sword. Despite its ability to achieve good learning for network fluctuations, it may also become a potential incentive to make the solutions not robust. Furthermore, the instability of ML algorithms may deteriorate the network fluctuations [252].
- **Scheme evaluation.** The experimental evaluations of most existing schemes are based on simulations lacking of verification in real network environment and the network scale is relatively small with simple topologies, which makes the experimental results less convincing.
- **Micro-burst tolerance.** Micro-burst is a common traffic pattern in modern data centers, which can exacerbate the problem of network congestion [122, 195, 225, 261]. Reasonable absorption of micro-burst traffic can effectively improve the overall robustness of the network, and performs better than adjusting the congestion window. Unfortunately, there are few ML-based congestion control schemes considering the mitigation of micro-burst, which is believed to be a valuable research topic and a good research opportunity [315].

All in all, the current research in this field is still quite limited, and the modeling, algorithm design, experimental method and applications are still in primary research phase, and there is still considerable research value in applying ML to the congestion control.

3.8 Fault Management

A typical data center is usually composed of hundreds or thousands of various devices, which are typically not 100% reliable with occasional failures. Thereby, fault management has become an important but complex O&M problem for a large

Table 3.12 Research progress of data center network intelligence: congestion control

Ref	Category[a]	ML category and model adopted	Features	Data source	Feature selection	Additional constraints	Estimation function	Experimental comparison subjects
Jin et al. [124]	C	RL, Q-learning and Sarsa	Applying improved classical reinforcement learning algorithms to software defined data center networks	Simulated data	Flow information, link information, etc.	Link utilization	Occupied bandwidth, average link utilization	✔|✔|✘
Liao et al. [145, 146]	D	DRL, Q-learning	A low latency and fast converging congestion avoidance strategy	Microsoft Research Cambridge Trace	Average node latency, data block read rate, etc.	None	Read/write latency, convergence speed, etc.	✔|✔|✔
Liu et al. [155, 157]	D	DRL, DQN	Reduce service latency with data placement policies	MSR Cambridge Traces [183]	End-to-end node information, read/write latency, subsequent analysis latency, etc.	None (As listed in Table 3.5, the additional constraint for this article is network latency)	Average data read latency, etc.	✔|✔|✔

(continued)

Table 3.12 (continued)

Ref	Category[a]	ML category and model adopted	Features	Data source	Feature selection	Additional constraints	Estimation function	Experimental comparison subjects
Majidi et al. [168]	D	DL, DNN	Dual queue independent dynamic threshold control based on flow classification for low latency and high throughput	Simulated data	Source IP, destination IP, source port, destination port, protocol information, etc.	High throughput and low losses	FCT, etc.	✓|✓|✗
Nie [185]	D	RL, UCB	Presented a TCP-RL system that dynamically configures information flows suitable for short flows via group-based RL and dynamically configures CC schemes suitable for long flows via deep RL	Baidu's Online Production Data Center	Transmission time, throughput, loss rate, RTT, etc.	Maximizing throughput and minimizing RTT	Response time, throughput, RTT, etc.	✓|✓|✓

Ruffy et al. [213]	C	DRL, DDPG etc.	A simulation simulator supporting different congestion control algorithms	Simulated data	Switch buffer occupancy, interface utilization, etc.	None	Bandwidth, etc.	✓|✓|✓
Thiruvenkatam et al. [250]	D	UL, K-means	Introduced two congestion control mechanisms to improve the transport protocol	Simulated data	Buffer size, dropped packets, etc.	None	Input throughput, etc.	✓|✓|✗
Sun et al. [239, 240]	C	DRL, DDPG	The DRL algorithm was used to improve power efficiency and ensure FCT. Moreover, the link margin ratio was adjusted according to the policy generated by DRL agent to deal with the congestion caused by unpredicted bursts	Wikipedia trace files [259]	Number of switch ports, port power, incoming and outgoing flow rates, etc.	QoS and energy consumption	FCT, etc.	✓|✓|✗

(continued)

Table 3.12 (continued)

Ref	Category[a]	ML category and model adopted	Features	Data source	Feature selection	Additional constraints	Estimation function	Experimental comparison subjects
Rastegarfar et al. [207]	C	DL, DNN	Congestion control through flow classification accuracy and optical bandwidth aggregation	Simulated data	5-tuple, packet size, etc.	None	Throughput, etc.	✓|✓|✗
Xiao et al. [283]	D	DRL, CNN/LSTM	Applied DRL to solve congestion control problems	Simulated data	congestion window size, RTT and inter-arrival time of ACKs	None	Average throughput, average RTT, etc.	✓|✓|✗

[a] For convenience, we use "C" for Centralized Congestion Control, and "D" for Distributed Congestion Control

scale data center network [27, 114, 196, 198, 221]. The current data center network O&M model is still passive and overly dependent on manual operations, which is inefficient and not intelligent. Thus, the new generation data center needs to draw support from new technologies to improve the O&M efficiency. Fortunately, artificial intelligence technology provides an effective way to tackle this challenging problem, and therefrom the ML-based intelligent O&M has naturally become a research hotspot in academia and industry. In the traditional data center reactive O&M model, fault prediction is mainly done by extracting the characteristics of system component messages [215] and correlation between logs and faults [276]. Salfner et al. [217] used Hidden Semi-markov Model (HSMM) to analyze message logs for fault prediction. Guo et al. [98] implemented end-to-end fault detection through TCP proxy. Herodotou et al. [111] utilized statistical data mining techniques for fault location, but their proposed probabilistic path model assumed only one link fault could occur, which may lead to anomalous fault reports. In the rest of this section, we will discuss and analyze the existing work with respect to different stages of the fault management life-cycle, including fault prediction, fault monitoring, fault location, and fault self-healing.

3.8.1 Fault Prediction

The large amount of alarm information, logs, and service impairment information generated in the network facilitate network failure prediction, and how to effectively extract critical information with fast sensing capability has been the primary concern of current research. Predicting the time period of the next failure and reacting to it in advance is deemed as the most feasible solution to improve data center reliability. As the key advantage of machine learning lies in prediction, thus numerous ML-based fault prediction algorithms have been put forward in recent years. Lee et al. argued that the prediction of failures should have the lowest possible computational latency [137]. They addressed the imbalance between normal and fault events through a two-stage fault prediction framework. The experimental results showed it achieved a 39.45% improvement in F_3-score [199] compared with multi-class SVM and logistic regression, and realized a fast fault prediction with the prediction time less than 9 ms, which could well meet the requirements of industrial data centers. Zhang et al. [308] took a different approach by incorporating data noise (irrelevant log information) into the optimization scheme and solving the imbalance problem by extracting the modularity of system log information. They observed that before the failure occurs the failure model of the same switch has some common syslog characteristics, which are further explored in this work based on a machine learning algorithm (i.e., Random Forest) to predict switch failures. Whereas, the above schemes have the same drawbacks as traditional statistical methods, which cannot deal with sequence data well in cloud data centers. Gao et al. [89] proposed a multilayer bidirectional LSTM based fault prediction model and adjusted the weights of input features by data classification, which has better applicability in cloud data centers.

3.8.2   Fault Detection

Fault detection is not only an important means to perceive network failures, but also one of the key technologies of fault management. Nevertheless, the large-scale nature of data center network and high richness of network data (e.g., log files, virtualization information, and regular network traffic) make fault detection extremely challenging in data centers. Bambharolia et al. [31] constructed a semi-supervised learning model to predict and detect faults, and subsequently employed one dimensionality reduction technique named Relevance Deduction to reduce the number of features in a large dataset. Garg et al. [91] improved the existing Gray Wolf Optimization algorithm (a meta-heuristic algorithm based on evolutionary computation, abbreviated as GWO) and CNN, and the improved GWO and improved CNN algorithms are responsible for the extraction of network features and the classification of network anomalies for fault detection, accordingly. The hybrid model achieved good results, realizing improvements of 3 to 8% in detection rate, false alarm rate, and accuracy compared to the standard GWO with CNN.

3.8.3   Fault Location

After a network fault has been detected, it is necessary to accurately locate the fault before adopting further fault self-healing or fault tolerance mechanisms. However, likewise fault detection, accurate fault localization is also a very complex and difficult problem, which has been proved to be an NPC problem [61]. Our investigations show that there is very little work in this field in recent years. The only existing research work we found is Yang et al. [290], which indicates that traditional non-ML fault location methods are typically limited by the search capability and usually fall into local optimal solutions, which have a loss in location accuracy. Thus, they introduced DNN to improve the global search capability and proposed an accurate fault location method applicable to large-scale alarm scenarios. Their improved fault propagation model and threshold mechanism are used to reduce the fault range and noise. Finally, the precise fault coordinates are derived from the sub-NN generated by cross-variance.

3.8.4   Fault Self-Healing

The ultimate goal of network O&M is to troubleshoot and maintain service continuity. Therefore, how to achieve rapid self-recovery after the occurrence of a failure and ensure that the service is not affected is the last but very important step of network O&M life cycle. Unfortunately, there are no existing relevant research works found yet, expect some preliminary attempts for fault self-healing.

Table 3.13 Assessment of fault management schemes based on REBEL-3S

Ref	Reliability	Energy efficiency	Bandwidth utilization	Latency	Security	Stability	Scalability
Garg et al. [91]	YES	NO	YES	NO	YES	YES	YES
Ganguly et al. [87]	YES	NO	YES	NO	NO	YES	YES
Yang et al. [290]	YES	NO	NO	YES	YES	YES	YES
Lee et al. [137]	YES	NO	YES	YES	NO	YES	YES
Marahatta [169, 170]	YES	YES	YES	NO	NO	YES	NO
Bambharolia et al. [31]	YES	NO	NO	NO	NO	YES	NO
Xie et al. [284]	YES	NO	YES	NO	NO	YES	YES
Liu et al. [151]	YES	NO	YES	NO	NO	YES	NO
Zhang et al. [308]	YES	NO	YES	NO	YES	YES	YES
Kimura et al. [132]	YES	NO	YES	NO	NO	YES	NO
Liu et al. [160]	YES	NO	NO	NO	NO	YES	YES
Gao et al. [89]	YES	NO	YES	YES	NO	YES	NO

For example, Zhang et al. [308] proposed a fault prevention mechanism to advance service reversal by predicting faults. Precisely, the authors predicted faults by using supervised learning-based models. When a fault is predicted to occur soon, they actively cut the link in advance and employ specific load balancing mechanisms to bypass the failure point and repair the failure point manually (Table 3.13).

3.8.5 Discussion and Insights

We present and compare the existing solutions with details under different categories in Table 3.14, and evaluate each solution from various aspects according to REBEL-3S in Table 3.13. Through the research, we summarize the general situation and critical points of ML in fault management. Supervised learning and deep learning are the primary models that are used in fault management, while reinforcement learning or deep reinforcement learning are less used.

- **False negative samples.** The false negative samples will seriously deteriorate the precisions of machine learning models. Taking fault prediction as an example, if the actual situation is a fault, but it is predicted to be normal, it will result in huge recovery cost to the network. Thus, the probability of false alarms should be minimized.
- **Data imbalance distribution.** The number of normal and fault events shows a high imbalance. According to Lee et al., only 8,957 out of every 104 million events were related to machine failure [137]. This proportion accounts for less than 1%, so the accuracy of intelligent solutions is difficult to be expressed intuitively by the prediction results.
- **Noise disturbances.** Fault management, especially the precise localization of faults, requires filtering irrelevant information to improve the ability of resistance

Table 3.14 Research progress of data center network intelligence: fault management

Ref	Category[a]	ML category and model adopted	Features	Data source	Feature selection	Additional constraints	Estimation function	Experimental comparison subjects
Garg et al. [91]	D	DL, CNN	Network anomaly detection and classification via GWO and CNN	Baseline datasets-DARPA'98 [1], KDD'99 [131] and synthetic datasets	5-tuple, etc.	Minimize feature set while reducing error rate	Accuracy, F_1-score, etc.	✓|✓|✓
Ganguly et al. [87]	P	SL, SVM	Predict disk failures through model training and disk scoring	SMART data, performance counters	Timestamps, repair detection times, disk IDs, etc.	None	Accuracy, recall, etc.	✓|✗|✗
Yang et al. [290]	L	DL, DNN	A precise fault location method for large-scale alarm scenarios was proposed	Network operator's management system and OpenStack system	Output optical power, input optical power, laser bias, etc.	Low computational latency	Positioning accuracy, training time, etc.	✓|✗|✓
Lee et al. [137]	P	SL, SVM/RF	A dual-stage framework for fault prediction was proposed	Google traces [94]	CPU usage, disk I/O time, disk space and memory usage, etc.	Low computational latency	F_3-score, AUC	✓|✗|✓
Marahatta [169, 170]	P	DL, DNN	Resource provisioning, fault prediction by task classification	Real data sets	CPU utilization, etc.	Energy overhead	Energy consumption etc.	✓|✓|✗

	P/D						None	No experiment
Bambharolia et al. [31]	P/D	Semi-supervised learning, DT	Predict and detect failures through pattern recognition of log messages between data centers	None	Server maintenance CPU usage, memory usage, etc.	None	None	✓|✓|✓
Xie et al. [284]	P	SL/UL, SVM/KNN, etc.	A Self-optimizing Modeling Engine for Disk Failure Prediction in Heterogeneous Data Centers	Backblaze dataset [22], Baidu dataset [313], CMRR dataset [181]	Disk model data	None	F1-score, etc.	✓|✗|✓
Liu et al. [151]	P	DL, DNN	An online fault prediction scheme based on ELM [8].	Google cluster traces [94]	Job priority, request resources, etc.	Low computational latency	Accuracy, etc.	✓|✗|✓
Zhang et al. [308]	P	SL, RF	Proactive failover by predictive model results	Real Data Sets	Status change logs of interfaces, links or neighbors, etc.	None	Accuracy, recall, etc.	✓|✗|✓
Kimura et al. [132]	D	SL, SVM	Fault detection based on structured log template	USENIX [47]	Log periodicity, log bursting, etc.	None	Accuracy, F_1-score, etc.	✓|✓|✗

(continued)

Table 3.14 (continued)

Ref	Category[a]	ML category and model adopted	Features	Data source	Feature selection	Additional constraints	Estimation function	Experimental comparison subjects
Liu et al. [160]	P	SL, RF	Applying random forest algorithms to fault prediction	Real data set	System event logs and machine check error logs, crash times, crash reasons, etc. for all machines	None	Accuracy, recall and warning time margins	✔|✔|✔
Gao et al. [89]	P	DL, LSTM	A multilayer bidirectional LSTM-based fault prediction model was proposed, and the weights of the input features were adjusted by data classification	Google cluster traces [49, 210]	CPU usage, memory usage, task priority, etc.	None	Accuracy, F_1-score, etc.	✔|✔|✔

[a] For convenience, we use "P" for Fault Prediction, "D" for Fault Detection, "L" for Fault Location, and "H" for Fault Self-healing

Table 3.15   Assessment of network security schemes based on REBEL-3S

Ref	Reliability	Energy efficiency	Bandwidth utilization	Latency	Security	Stability	Scalability
Zeng et al. [302]	YES	NO	YES	NO	YES	YES	NO
Schueller et al. [222]	YES	YES	YES	YES	YES	YES	YES
Garg et al. [91]	YES	YES	YES	NO	YES	YES	YES
Zekri et al. [301]	YES	YES	NO	YES	YES	YES	YES
Xiao et al. [280]	YES	YES	NO	NO	YES	YES	YES
Satheesh et al. [219]	YES	YES	YES	YES	YES	YES	YES
Abubakar et al. [3]	YES	YES	YES	NO	YES	YES	YES
Chen et al. [56]	YES	NO	YES	NO	YES	YES	YES
Baek et al. [23]	YES	NO	NO	NO	YES	YES	YES
Chen et al. [54]	YES	NO	YES	YES	YES	YES	YES

to interference [191, 204]. Typically, a single fault may lead to a chain reaction that causes multiple devices to generate alarm messages. The number of alarm messages provides little help to the alarm's effectiveness, on the contrary, false alarm messages can even reduce the accuracy of fault localization.

- **Complex dynamic environments.** Inevitably, there is diversity in management methods and data among different vendors and devices in large data center systems, and many ML methods cannot perform well for heterogeneous device data sets.

All in all, the current research in this field is still in its infancy, and the number of existing work is comparatively small. Moreover, most of these existing approaches only consider single fault scenario and lack effective means for fault root cause analysis and fault self-healing, resulting in that the closed-loop system of end to end intelligent O&M has not yet been formed in academia (Table 3.15).

3.9   Network Security

As cloud tenants pay more and more attention to user privacy and data security, cloud DC network security has become a key issue in the field of cloud computing. However, the complexity, heterogeneity and dynamics of cloud data center network pose great challenges to network security. The vulnerabilities in the network not only can cause leakage of users' private data, but also may be exploited by some malicious attackers, who will invade the network infrastructure through distributed denial of service (DDoS) and other means, resulting in massive disruption of network service. In the Uptime Institute's 2020 Data Center Industry Survey [2], security was highlighted as one of the key influences driving demand for enterprise data centers. According to the investigations of Cisco visual network index (VNI) [201], there have been nearly 17 million DDoS intrusions by 2020, which is three times of 2015. In response to this critical issue, many researchers and scholars have begun to explore the security protection mechanisms of data center network, as well as the design scheme of security system such as Network Intrusion Detection Systems (NIDS). Kawahara et al. [130] proposed a flow statistics-based IDS scheme. Mai et al. [166] proposed a detection algorithm for non-volume-dependent anomalies based on anomaly statistics and sampling. Xiang et al. [279] used generic entropy and information distance to detect low-speed DDoS attacks, but the scheme requires the control of all routers. Although these traditional schemes, to some extent, have made some improvements to the data center network security, they still suffer from low accuracy, high false alarm rate, and high memory consumption. Coupled with the heterogeneity and diversity of DC network environment, the traditional techniques will no longer be suitable for data center network.

Recently, the machine learning based network anomaly detection technology has attracted extensive attention of researchers [238]. Empirically, machine learning can be effectively used in the design of NIDS to improve the detection accuracy [90] and achieve low false alarm rate [173]. Previous research efforts have conducted a rich and detailed investigation but have not explored for DCN intelligent scenarios [42, 90, 256]. In the work [280], Xiao et al. proposed a CKNN-based DDoS attack detection method and a grid-based low-cost data reduction method, which not only improves the classification accuracy but also reduces training cost by exploring the correlation between training data. In the next three years, supervised learning started to be gradually applied to the network security in DCN scenarios. Zekri et al. [301] combined C4.5 algorithm and signature detection techniques to achieve automatic and effective detection of signature attacks. Baek et al. [23] performed anomaly detection by using a supervised machine learning algorithm based on assumed clustering labels. Schueller et al. [222] proposed a lightweight and scalable hierarchical Intrusion Detection System (IDS) architecture based on support vector machine (SVM). In recent few years, with the rise of deep learning, researchers have also begun to explore a new application paradigm of deep learning in the field of data center network security. Zeng et al. [302] proposed a lightweight end-to-end framework for flow classification and intrusion detection based on DL with excellent performance on two public datasets. Garg et al. [91] performed network anomaly detection and classification by GWO and CNN.

Our research findings together with some key features of the investigated intelligent solutions are summarized in Table 3.14, and the Table 3.13 lists the evaluation results of each solution according to the REBEL-3S criteria. Additionally, we also put forward some insights and observations for reference, as listed below (Table 3.16).

- **Distributed detection.** Currently, the exiting security detection schemes mainly work in a centralized way, which usually results in a high cost in both network monitoring and intrusion detection [257]. Therefore, most of the existing solutions consider the monitoring overhead in their optimization models. To solve the high-cost problem of centralized solution thoroughly, we contend that it would be preferable to adopt a distributed detection scheme so as to relieve the pressure on a single centralized controller and avoid the single point of failure

Table 3.16 Research progress of data center network intelligence: network security

Ref	ML category and model adopted	Features	Data source	Feature selection	Additional constraints	Estimation function	Experimental comparison subjects
Zeng et al. [302]	DL, CNN/LSTM	A lightweight end-to-end framework for flow classification and intrusion detection was proposed, and it had excellent performance on two public data sets	ISCX 2012 IDS dataset [231] respectively and ISCX VPN-nonVPN flow dataset [72]	Unspecified	Storage	F_1-score, etc.	✗|✗|✓
Schueller et al. [222]	UL, SVM	Lightweight and scalable hierarchical IDS architecture	MIT Lincoln Laboratory's DARPA dataset [67]	Stream characteristics such as average number of packets per stream, average number of bytes per stream, average duration of stream	Network overhead	Detection rate, FAR, etc.	✓|✓|✗
Garg et al. [91]	DL, CNN	Network anomaly detection and classification via GWO and CNN	Baseline datasets–DARPA'98 [1], KDD'99 [131] and synthetic datasets	5-tuple, etc.	Minimize feature set while reducing error rate	Accuracy, F_1-score, etc.	✓|✓|✓
Zekri et al. [301]	SL, C4.5	Combine C4.5 algorithm and signature detection technology to achieve automatic and effective detection of signature attacks	Simulated data	Protocol, service, TTL, etc.	Network overhead, detection rate, etc.	DDoS attack detection time, recall, etc.	✓|✗|✓

(continued)

Table 3.16 (continued)

Ref	ML category and model adopted	Features	Data source	Feature selection	Additional constraints	Estimation function	Experimental comparison subjects
Xiao et al. [280]	UL, KNN	Proposed a CKNN-based DDoS attack detection method and a low-cost data reduction method based on the grid	KDD'99 data set [116] and two other real data sets	5-tuple, etc.	Network overhead	Time, Accuracy, etc.	✓|✗|✓
Satheesh et al. [219]	SL, DT	Anomaly intrusion detection based on flow classification and prioritization	NSL-KDD dataset [263]	Ethernet source, ingress port and destination, VLAN ID, priority, etc.	Network overhead, QoS	MAE, Recall, etc.	✓|✗|✓
Abubakar et al. [3]	DL, KNN	Integration of pattern recognition-based NN models into IDS	NSL-KDD dataset [117]	Unspecified	Network overhead	Accuracy, F_1-score, etc.	✓|✗|✓
Chen et al. [56]	SL/UL/DL, RF/K-Means, etc.	Anomaly detection framework focusing on single metric and multi-metric relationships	Real metadata from the Lustre file system [71] in the IHEP data center	Range of statistical features and fitted features of time series	None	ME, MAE, RMSE, etc.	✓|✓|✓
Baek et al. [23]	SL, SVM/RF, etc.	Anomaly detection by assumed clustering labels for supervised learning	NSL-KDD dataset [245]	Similar to the way described in [197]	None	Accuracy, etc.	✓|✗|✗
Chen et al. [54]	SL, GBDT	Anomaly detection by network flow classification	NSL-KDD dataset [116]	Basic TCP connection characteristics, TCP connection content characteristics and time-based network flow statistics characteristics	Detection speed	Accuracy, FPR, etc.	✓|✗|✓

problem as well. Admittedly, there are also some key concerns to be addressed in distributed solutions, such as the relatively high complexity of distributed system, the consistency problem, and how to avoid falling into local optimization, etc.

- **Computing Overhead.** In these ML-based intelligent security schemes, ML techniques are usually leveraged to distinguish network attacks based on flow characteristics, where data acquisition and statistics are necessarily required. However, the constant data acquisition and processing will potentially bring considerable computing overhead to the controller in a large and complex DC network [92]. Especially for persistent network attacks, the responsiveness of the controller is further degraded. Thus, how to mitigate the computing overhead introduced by ML related operations still remains an open problem to be addressed.

- **False alarm rates.** Ideally, ML algorithms must avoid high false alarm rates. A high false alarm rate can cause a waste of network resources and seriously affect the overall network performance. Therefore, the selection and design of ML algorithms used in security schemes need to take the false alarm rate into account.

3.10 New Intelligent Networking Concepts

With the booming development of ML applications in the fields mentioned above, the data center network design is enabled to realize some functions that were impossible before, thereby the data center network continues to evolve from the network softwarization to the network intelligentization. According to our investigation, most of the existing intelligent solutions adopt SDN to achieve a global network control and optimization, where the ML-based intelligence modules are centrally implemented in the SDN controller. However, many hardware vendors contradict SDN's original intention and try to establish a closed system through vendor lock-in, which will inhibit the development of network intelligence in the long term. Besides, the performance and scalability problem of SDN controllers will also hinder the promotion of network intelligence to a certain extent. Additionally, the current multi-controller mechanisms (e.g. dual-master or master-standby) targeting at avoiding the single point of failure of controllers still have problems in real-time failing over, state synchronization and policy consistency. To this end, on the one hand, both the academia and industry are making efforts to further study and improve the ML-assisted SDN technologies, on the other hand, they are also trying new technology development breakthrough direction, and put forward a series of new intelligent network systems, which push the network intelligence to a new stage. In this section, we will briefly introduce some newly proposed representative intelligent networking concepts, as listed in Table 3.17.

Table 3.17 List of new intelligent networking concepts

Originator	New networking concepts	Year of release
David Lenrow et al. [139]	Intent-driven network (IDN)	2015
Mestres et al. [175]	Knowledge-defined network (KDN)	2017
Juniper [249]	Self-driving network (SDN)	2017
Gartner [140]	Intent-based network (IBN)	2017
Cisco [118]	Intent-based network (IBN)	2017
Huawei [120]	Intent-driven network (IDN)	2018

3.10.1 Intent-Driven Network

The basic idea of IDN was first revealed in a draft by David Lenrow [139], chair of the Open Networking Foundation's Northbound Interface Working Group. Likewise, Intent-driven network (IDN) is a self-driven network as well. An intent implies the desired network properties and functionalities that we want the network to offer, and can be described as high-level semantic statements of their macro-level behaviors. IDN aims to automate such intents of applications by utilizing a closed-loop system together with the decoupled network control logic. Drawing support from artificial intelligence technology, it realizes the transformation from passive network O&M to intelligent active O&M, the predictability of the network, early identification of network faults, and active network optimization according to application intents.

Nevertheless, IDNs are still facing severe challenges [190], such as continuous closed-loop verification and automated deployment. Furthermore, a unified and precise definition of IDN has not yet been formed. In practical terms, Gartner proposed an intent-based networking system (IBNS), Cisco launched an implementation called Intent-based Network (IBN), and Huawei launched a program directly called IDN.

3.10.2 Knowledge-Defined Network

The KDN was initially inspired by the concept proposed by Clark et al. in 2003 [60] and finally formalized by Mestres et al. in 2017 [175]. Unlike the SDN with a flat structure, the KDN can be simply summarized as a fusion of SDN with network analysis and artificial intelligence.

As shown in Fig. 3.3, the KDN has four main components, namely Data Plane (DP), Control Plane (CP), Management Plane (MP), and Knowledge Plane (KP). The main functions of DP and CP are similar to the corresponding modules (data plane and control plane) in SDN. MP has three prominent responsibilities. First, to ensure reliable network operation and performance meeting the specific requirements of network topology and network device configuration. Comparatively, in

Fig. 3.3 KDN planes

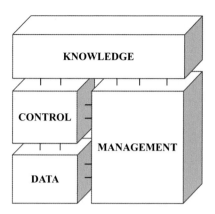

SDN paradigm this functionality is implemented in the control plane. Second, to provide efficient network monitoring and network analysis. Third, to collect historical data on demand in due course of time. The KP implements decision-oriented behavioral models and reasoning mechanisms by employing deep learning to transform the collected historical data into knowledge (useful information), which is further utilized to make decisions.

3.10.3   Self-Driving Network

As the field of self-driving vehicles surges forward, the network researchers believe that what they have learned from the successful experience of self-driving vehicles can be applied to the network to realize an automatic driving network. With this aim, Juniper proposed the Self-Driving Network (SDN) [249]. Juniper regarded the Self-Driving Network as an autonomous network that can predict and adapt to dynamic environments to implement network configuration, monitoring, management, and security defense with little or even no human intervention. With the Self-Driving Network, the network performance problems can be anticipated before they affect the user experience. The implementation of Self-Driving Network requires three main automation strategies, as described below.

1. Reduce operational complexity by simplifying and abstracting the network.
2. Accelerate the deployment of network services.
3. Improve network resource utilization and network resiliency using deep remote sensing techniques.

Moreover, the above automation strategies require the technical support of telemetry [128], automation, ML [18] and declarative programming. Inspired by the concept of Self-Driving Network, Huawei has also proposed a similar new concept, named Autonomous Driving Network (ADN), which is though implemented in a

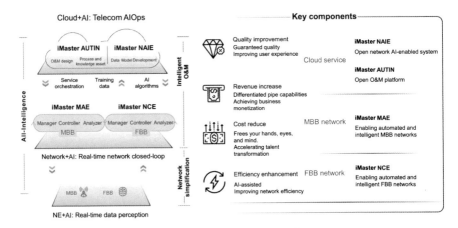

Fig. 3.4 Huawei ADN system panorama. Image from *Huawei ADN Solution White Paper*

different way, but has similarity in philosophy. According to the *Huawei ADN Solution White Paper* released by Huawei in 2020 [4], the core idea of ADN is similar to Self-Driving Network but it provides more specific details about implementations and has launched commercial ADN products. The panorama of Huawei's ADN system is shown in Fig. 3.4.

3.10.4 Intent-Based Network (Gartner)

Since the concept of IDN was introduced in 2015, Gartner also released a report [140] in early 2017, which defined intent-based networking (IBN) and proposed a corresponding systematized network system called IBNS. The IBN is an intelligent network scheme that monitors overall network performance, identifies problems and resolves them automatically without human intervention. The workflow of IBNS is as depicted in Fig. 3.5. It is worth mentioning that Software Defined Network and IBNS can be deployed independently or jointly in the network to achieve the best overall benefit. Gartner's definition of IBN includes the following four key points.

1. Translation and validation: The system takes higher-level services policies from the end-user as input and translates them into the executable network configurations. The system then generates and validates the correctness of the generated design and configurations.
2. Automated Implementation: Through network automation and network orchestration, the system can automatically make appropriate network changes.
3. Network state awareness: The system provides real-time network states for the systems it manages and controls.

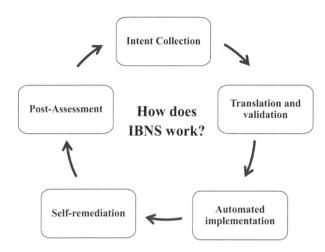

Fig. 3.5 IBNS workflow

4. Intent Assured Optimization: The system continuously verifies whether the original intents are satisfied, and take remedial measures whenever the desired intents are distorted.

In short, IBN allows network administrators to define the network as what they want and have an automated network management system (i.e., IBNS) to enforce policies. At present, many startups have started their services in IBN, including Apstra, Forward Networks, Waltz, Veriflow, and so on. More and more researchers also begin to devote their efforts to the research in IBNS [43, 105, 216]. Overall, the IBN is a nascent technology with a beautiful vision, but some of the technologies involved are still in infancy and need to be further improved and validated in practice.

3.10.5 *Intent-Based Network (Cisco)*

In June 2017, Cisco released a solution called Intent-Based Network [119], claiming that "IBN is the network that will shape the next 30 years". Meanwhile, Cisco has also released software and hardware to build IBNs, such as Software-Defined Access [235] and Catalyst 9000 switching groups [59]. Cisco insists that the rise of artificial intelligence can empower the network with a higher level of automation capabilities and bridge the gap between the original service intents and the final executed network policies. Specifically, IBN offers five benefits: better service agility; higher operational efficiency; continuous alignment of the network with service objectives; better compliance and security; and lower risk.

References

1. 1998 DARPA intrusion detection evaluation dataset (1998). https://www.ll.mit.edu/r-d/datasets/1998-darpa-intrusion-detection-evaluation-dataset
2. 2020 data center industry survey - uptime institute (2020). https://uptimeinstitute.com/resources/asset/2020-data-center-industry-survey
3. A. Abubakar, B. Pranggono, Machine learning based intrusion detection system for software defined networks, in *2017 Seventh International Conference on Emerging Security Technologies (EST)* (IEEE, Piscataway, 2017), pp. 138–143
4. ADN-solution-white-paper (2020). https://www-ctc.huawei.com/en/news/2020/5/adn-solution-white-paper
5. A. Aguado, P.A. Haigh, E. Hugues-Salas, R. Nejabati, D. Simeonidou, Towards a control plane management architecture enabling proactive network predictability, in *2016 Optical Fiber Communications Conference and Exhibition (OFC)* (IEEE, Piscataway, 2016), pp. 1–3
6. M. Aibin, Traffic prediction based on machine learning for elastic optical networks. Opt. Switch. Netw. **30**, 33–39 (2018)
7. M. Aibin, K. Walkowiak, S. Haeri, L. Trajković, Traffic prediction for inter-data center cross-stratum optimization problems, in *2018 International Conference on Computing, Networking and Communications (ICNC)* (IEEE, Piscataway, 2018), pp. 393–398
8. A. Akusok, K.M. Björk, Y. Miche, A. Lendasse, High-performance extreme learning machines: a complete toolbox for big data applications. IEEE Access **3**, 1011–1025 (2015)
9. M. Al-Fares, S. Radhakrishnan, B. Raghavan, N. Huang, A. Vahdat, et al., Hedera: dynamic flow scheduling for data center networks, in *NSDI'10: Proceedings of the 7th USENIX Conference on Networked Systems Design and Implementation, San Jose*, vol. 10, pp. 89–92
10. Alibaba/clusterdata (2017). https://github.com/alibaba/clusterdata
11. M. Alizadeh, A. Greenberg, D.A. Maltz, J. Padhye, P. Patel, B. Prabhakar, S. Sengupta, M. Sridharan, Data center TCP (DCTCP), in *Proceedings of the ACM SIGCOMM 2010 Conference* (2010), pp. 63–74
12. M. Alizadeh, A. Kabbani, T. Edsall, B. Prabhakar, A. Vahdat, M. Yasuda, Less is more: trading a little bandwidth for ultra-low latency in the data center, in *9th USENIX Symposium on Networked Systems Design and Implementation (NSDI'12)* (2012), pp. 253–266
13. M. Alizadeh, S. Yang, M. Sharif, S. Katti, N. McKeown, B. Prabhakar, S. Shenker, pfabric: minimal near-optimal datacenter transport. ACM SIGCOMM Comput. Commun. Rev. **43**(4), 435–446 (2013)
14. M. Alizadeh, T. Edsall, S. Dharmapurikar, R. Vaidyanathan, K. Chu, A. Fingerhut, V.T. Lam, F. Matus, R. Pan, N. Yadav, et al., Conga: distributed congestion-aware load balancing for datacenters, in *Proceedings of the 2014 ACM Conference on SIGCOMM* (2014), pp. 503–514
15. P. Amaral, J. Dinis, P. Pinto, L. Bernardo, J. Tavares, H.S. Mamede, Machine learning in software defined networks: data collection and traffic classification, in *2016 IEEE 24th International Conference on Network Protocols (ICNP)* (IEEE, Piscataway, 2016), pp. 1–5
16. F. Amezquita-Suarez, F. Estrada-Solano, N.L. da Fonseca, O.M.C. Rendon, An efficient mice flow routing algorithm for data centers based on software-defined networking, in *ICC 2019-2019 IEEE International Conference on Communications (ICC)* (IEEE, Piscataway, 2019), pp. 1–6
17. T.W. Anderson, The statistical analysis of time series. John Wiley & Sons, 2011.
18. AppFormix overview - TechLibrary - juniper networks (2018). https://www.juniper.net/documentation/en_US/appformix/topics/concept/about-appformix.html
19. Application layer packet classifier for linux (2008). http://l7-filter.sourceforge.net/
20. M. Arlitt, T. Jin, A workload characterization study of the 1998 world cup web site. IEEE Netw. **14**(3), 30–37 (2000)

21. J. Athavale, Y. Joshi, M. Yoda, Artificial neural network based prediction of temperature and flow profile in data centers, in *2018 17th IEEE Intersociety Conference on Thermal and Thermomechanical Phenomena in Electronic Systems (ITherm)* (IEEE, Piscataway, 2018), pp. 871–880

22. Backblaze hard drive stats (2013). https://www.backblaze.com/b2/hard-drive-test-data.html

23. S. Baek, D. Kwon, J. Kim, S.C. Suh, H. Kim, I. Kim, Unsupervised labeling for supervised anomaly detection in enterprise and cloud networks, in 2017 IEEE 4th International Conference on Cyber Security and Cloud Computing (CSCloud) (IEEE, Piscataway, 2017), pp. 205–210

24. A. Baer, P. Casas, A. D'Alconzo, P. Fiadino, L. Golab, M. Mellia, E. Schikuta, Dbstream: a holistic approach to large-scale network traffic monitoring and analysis. Comput. Netw. **107**, 5–19 (2016)

25. W. Bai, L. Chen, K. Chen, H. Wu, Enabling ECN in multi-service multi-queue data centers, in *13th USENIX Symposium on Networked Systems Design and Implementation (NSDI'16)* (2016), pp. 537–549

26. W. Bai, L. Chen, K. Chen, D. Han, C. Tian, H. Wang, PIAS: practical information-agnostic flow scheduling for commodity data centers. IEEE/ACM Trans. Netw. **25**(4), 1954–1967 (2017)

27. L.N. Bairavasundaram, G.R. Goodson, S. Pasupathy, J. Schindler, An analysis of latent sector errors in disk drives, in *Proceedings of the 2007 ACM SIGMETRICS International Conference on Measurement and Modeling of Computer Systems* (2007), pp. 289–300

28. M. Balanici, S. Pachnicke, Machine learning-based traffic prediction for optical switching resource allocation in hybrid intra-data center networks, in *Optical Fiber Communication Conference, Optical Society of America* (2019), pp. Th1H–4

29. M. Balanici, S. Pachnicke, Multi-step forecasting of intense traffic streams using machine learning for optical circuit switching, in *2019 21st International Conference on Transparent Optical Networks (ICTON)* (IEEE, Piscataway, 2019), pp. 1–4

30. M. Balanici, S. Pachnicke, Server traffic prediction using machine learning for optical circuit switching scheduling, in *Photonic Networks; 20th ITG-Symposium, VDE* (2019), pp. 1–3

31. P. Bambharolia, P. Bhavsar, V. Prasad, Failure prediction and detection in cloud datacenters. Int. J. Sci. Technol. Res. **6**(9), 1–6 (2017)

32. T. Benson, A. Akella, Data set for IMC 2010 data center measurement (2010). https://pages.cs.wisc.edu/~tbenson/IMC10_Data.html

33. T. Benson, A. Akella, D.A. Maltz, Network traffic characteristics of data centers in the wild, in *Proceedings of the 10th ACM SIGCOMM Conference on Internet Measurement* (2010), pp. 267–280

34. L. Bernaille, R. Teixeira, Implementation issues of early application identification, in *Asian Internet Engineering Conference* (Springer, Berlin, 2007), pp. 156–166

35. L. Bernaille, R. Teixeira, I. Akodkenou, A. Soule, K. Salamatian, Traffic classification on the fly. ACM SIGCOMM Comput. Commun. Rev. **36**(2), 23–26 (2006)

36. K.B. Bey, F. Benhammadi, F. Sebbak, M. Mataoui, New tasks scheduling strategy for resources allocation in cloud computing environment, in *2015 6th International Conference on Modeling, Simulation, and Applied Optimization (ICMSAO)* (IEEE, Piscataway, 2015), pp. 1–5

37. J.D.M. Bezerra, A.J. Pinheiro, C.P. de Souza, D.R. Campelo, Performance evaluation of elephant flow predictors in data center networking. Future Gen. Comput. Syst. **102**, 952–964 (2020)

38. T. Bhatia, Thomasbhatia/OpenDPI (2018). https://github.com/thomasbhatia/OpenDPI

39. I. Bolodurina, D. Parfenov, Model control of traffic by using data flows classification of the cloud applications in software-defined infrastructure of virtual data center, in *2017 40th International Conference on Telecommunications and Signal Processing (TSP)* (IEEE, Piscataway, 2017), pp. 8–11

40. I. Bolodurina, D. Parfenov, Comprehensive approach for optimization traffic routing and using network resources in a virtual data center. Proc. Comput. Sci. **136**, 62–71 (2018)

41. R. Boutaba, M.A. Salahuddin, N. Limam, S. Ayoubi, N. Shahriar, F. Estrada-Solano, O.M. Caicedo, A comprehensive survey on machine learning for networking: evolution, applications and research opportunities. J. Internet Services Appl. **9**(1), 1–99 (2018)

42. A.L. Buczak, E. Guven, A survey of data mining and machine learning methods for cyber security intrusion detection. IEEE Commun. Surv. Tuts. **18**(2), 1153–1176 (2015)

43. A. Campanella, Intent based network operations, in *2019 Optical Fiber Communications Conference and Exhibition (OFC)* (IEEE, Piscataway, 2019), pp. 1–3

44. X. Cao, Y. Zhong, Y. Zhou, J. Wang, C. Zhu, W. Zhang, Interactive temporal recurrent convolution network for traffic prediction in data centers. IEEE Access **6**, 5276–5289 (2017)

45. L. Caviglione, M. Gaggero, M. Paolucci, R. Ronco, Deep reinforcement learning for multi-objective placement of virtual machines in cloud datacenters. Soft Comput. **25**, 12569–12588 (2020)

46. C.C.f.A.I.D. Analysis, CAIDA data - overview of datasets, monitors, and reports (2020). https://www.caida.org/data/overview/index.xml

47. CFDR data (2005). https://www.usenix.org/cfdr-data

48. H. Che, Z. Bai, R. Zuo, H. Li, A deep reinforcement learning approach to the optimization of data center task scheduling. Complexity **2020**, 3046769 (2020)

49. X. Chen, C.D. Lu, K. Pattabiraman, Failure analysis of jobs in compute clouds: a google cluster case study, in *2014 IEEE 25th International Symposium on Software Reliability Engineering* (IEEE, Piscataway, 2014), pp. 167–177

50. R. Chen, C.Y. Liang, W.C. Hong, D.X. Gu, Forecasting holiday daily tourist flow based on seasonal support vector regression with adaptive genetic algorithm. Appl. Soft Comput. **26**, 435–443 (2015)

51. L. Chen, K. Chen, W. Bai, M. Alizadeh, Scheduling mix-flows in commodity datacenters with karuna, in *Proceedings of the 2016 ACM SIGCOMM Conference* (2016), pp. 174–187

52. X. Chen, J. Guo, Z. Zhu, R. Proietti, A. Castro, S.B. Yoom, Deep-RMSA: a deep-reinforcement-learning routing, modulation and spectrum assignment agent for elastic optical networks, in *2018 Optical Fiber Communications Conference and Exposition (OFC)* (IEEE, Piscataway, 2018), pp. 1–3

53. X. Chen, R. Proietti, H. Lu, A. Castro, S.B. Yoo, Knowledge-based autonomous service provisioning in multi-domain elastic optical networks. IEEE Commun. Mag. **56**(8), 152–158 (2018)

54. Z. Chen, F. Jiang, Y. Cheng, X. Gu, W. Liu, J. Peng, XGBoost classifier for DDoS attack detection and analysis in SDN-based cloud, in *2018 IEEE International Conference on Big Data and Smart Computing (bigcomp)* (IEEE, Piscataway, 2018), pp. 251–256

55. Z. Chen, J. Hu, G. Min, Learning-based resource allocation in cloud data center using advantage actor-critic, in *ICC 2019-2019 IEEE International Conference on Communications (ICC)* (IEEE, Piscataway, 2019), pp. 1–6

56. J. Chen, L. Wang, Q. Hu, Machine learning-based anomaly detection of ganglia monitoring data in hep data center, in *EPJ Web of Conferences, EDP Sciences*, vol. 245 (2020), p. 07061

57. I. Cho, K. Jang, D. Han, Credit-scheduled delay-bounded congestion control for datacenters, in *Proceedings of the Conference of the ACM Special Interest Group on Data Communication* (2017), pp. 239–252

58. A.K. Choudhury, E.L. Hahne, Dynamic queue length thresholds for shared-memory packet switches. IEEE/ACM Trans. Netw. **6**(2), 130–140 (1998)

59. Cisco catalyst 9000 wireless and switching family portfolio (2017). https://www.cisco.com/c/en/us/solutions/enterprise-networks/catalyst-9000.html

60. D.D. Clark, C. Partridge, J.C. Ramming, J.T. Wroclawski, A knowledge plane for the internet, in *Proceedings of the 2003 Conference on Applications, Technologies, Architectures, and Protocols for Computer Communications* (2003), pp. 3–10

61. G.F. Cooper, The computational complexity of probabilistic inference using Bayesian belief networks. Artif. Intell. **42**(2–3), 393–405 (1990)

62. B. Cronkite-Ratcliff, A. Bergman, S. Vargaftik, M. Ravi, N. McKeown, I. Abraham, I. Keslassy, Virtualized congestion control, in *Proceedings of the 2016 ACM SIGCOMM Conference* (2016), pp. 230–243

63. A.R. Curtis, W. Kim, P. Yalagandula, Mahout: low-overhead datacenter traffic management using end-host-based elephant detection, in *2011 Proceedings IEEE INFOCOM* (IEEE, Piscataway, 2011), pp. 1629–1637

64. A. Cuzzocrea, E. Mumolot, P. Corona, Coarse-grained workload categorization in virtual environments using the dempster-shafer fusion, in 2015 IEEE 19th International Conference on Computer Supported Cooperative Work in Design (CSCWD) (IEEE, Piscataway, 2015), pp. 472–477

65. A. Dainotti, A. Pescape, K.C. Claffy, Issues and future directions in traffic classification. IEEE Netw. **26**(1), 35–40 (2012)

66. Data miner 2 - settlements verified hourly LMPs (2011). https://dataminer2.pjm.com/feed/rt_da_monthly_lmps

67. Datasets | MIT lincoln laboratory (2016). https://www.ll.mit.edu/r-d/datasets?keywords=DARPA

68. J. Dean, S. Ghemawat, Mapreduce: simplified data processing on large clusters. Commun. ACM **51**(1), 107–113 (2008)

69. A.R. Doke, K. Sangeeta, Deep reinforcement learning based load balancing policy for balancing network traffic in datacenter environment, in *2018 Second International Conference on Green Computing and Internet of Things (ICGCIoT)* (IEEE, Piscataway, 2018), pp. 1–5

70. M. Dong, T. Meng, D. Zarchy, E. Arslan, Y. Gilad, B. Godfrey, M. Schapira, PCC vivace: online-learning congestion control, in *15th USENIX Symposium on Networked Systems Design and Implementation (NSDI'18)* (2018), pp. 343–356

71. S. Donovan, G. Huizenga, A.J. Hutton, C.C. Ross, M.K. Petersen, P. Schwan, Lustre: building a file system for 1000-node clusters, in *Proceedings of the Linux Symposium*, vol. 2003 (2003)

72. G. Draper-Gil, A.H. Lashkari, M.S.I. Mamun, A.A. Ghorbani, Characterization of encrypted and VPN traffic using time-related, in *Proceedings of the 2nd International Conference on Information Systems Security and Privacy (ICISSP)* (2016), pp. 407–414

73. C. Draxl, A. Clifton, B.M. Hodge, J. McCaa, The wind integration national dataset (wind) toolkit. Appl. Energy **151**, 355–366 (2015)

74. A. Duque-Torres, F. Amezquita-Suárez, O.M. Caicedo Rendon, A. Ordóñez, W.Y. Campo, An approach based on knowledge-defined networking for identifying heavy-hitter flows in data center networks. Appl. Sci. **9**(22), 4808 (2019)

75. N. Elprince, Autonomous resource provision in virtual data centers, in *2013 IFIP/IEEE International Symposium on Integrated Network Management (IM 2013)* (IEEE, Piscataway, 2013), pp. 1365–1371

76. J. Erman, M. Arlitt, A. Mahanti, Traffic classification using clustering algorithms, in *Proceedings of the 2006 SIGCOMM Workshop on Mining Network Data* (2006), pp. 281–286

77. J. Erman, A. Mahanti, M. Arlitt, C. Williamson, Identifying and discriminating between web and peer-to-peer traffic in the network core, in *Proceedings of the 16th international conference on World Wide Web* (2007), pp. 883–892

78. F. Estrada-Solano, O.M. Caicedo, N.L. Da Fonseca, Nelly: flow detection using incremental learning at the server side of SDN-based data centers. IEEE Trans. Ind. Inf. **16**(2), 1362–1372 (2019)

79. Q. Fang, Z. Li, Y. Wang, M. Song, J. Wang, A neural-network enhanced modeling method for real-time evaluation of the temperature distribution in a data center. Neural Comput. Appl. **31**(12), 8379–8391 (2019)

80. H. Feng, Y. Shu, Study on network traffic prediction techniques, in *Proceedings of 2005 International Conference on Wireless Communications, Networking and Mobile Computing, 2005*, vol. 2 (IEEE, Piscataway, 2005), pp. 1041–1044

81. J. Fiala, A survey of machine learning applications to cloud computing (2015). http://www.cse.wustl.edu/jain/cse57015/ftp/cld_ml/index.html

82. M. Finsterbusch, C. Richter, E. Rocha, J.A. Muller, K. Hanssgen, A survey of payload-based traffic classification approaches. IEEE Commun. Surv. Tuts. **16**(2), 1135–1156 (2013)
83. T. Flach, N. Dukkipati, A. Terzis, B. Raghavan, N. Cardwell, Y. Cheng, A. Jain, S. Hao, E. Katz-Bassett, R. Govindan, Reducing web latency: the virtue of gentle aggression, in *Proceedings of the ACM SIGCOMM 2013 Conference on SIGCOMM* (2013), pp. 159–170
84. F. Francois, E. Gelenbe, Optimizing secure SDN-enabled inter-data centre overlay networks through cognitive routing, in *2016 IEEE 24th International Symposium on Modeling, Analysis and Simulation of Computer and Telecommunication Systems (MASCOTS)* (IEEE, Piscataway, 2016), pp. 283–288
85. Z. Fu, Z. Liu, J. Li, Efficient parallelization of regular expression matching for deep inspection, in *2017 26th International Conference on Computer Communication and Networks (ICCCN)* (IEEE, Piscataway, 2017), pp 1–9
86. Q. Fu, E. Sun, K. Meng, M. Li, Y. Zhang, Deep q-learning for routing schemes in SDN-based data center networks. IEEE Access **8**, 103491–103499 (2020)
87. S. Ganguly, A. Consul, A. Khan, B. Bussone, J. Richards, A. Miguel, A practical approach to hard disk failure prediction in cloud platforms: big data model for failure management in datacenters, in 2016 IEEE Second International Conference on Big Data Computing Service and Applications (BigDataService) (IEEE, Piscataway, 2016), pp. 105–116
88. C. Gao, V.C. Lee, K. Li, DemePro: decouple packet marking from enqueuing for multiple services with proactive congestion control. IEEE Trans. Cloud Comput. **99**, 1–14 (2017)
89. J. Gao, H. Wang and H. Shen, Task failure prediction in cloud data centers using deep learning, in *IEEE Transactions on Services Computing*, vol. 15, no. 3, pp. 1411–1422, (2022). https://doi.org/10.1109/TSC.2020.2993728
90. P. Garcia-Teodoro, J. Diaz-Verdejo, G. Maciá-Fernández, E. Vázquez, Anomaly-based network intrusion detection: techniques, systems and challenges. Comput. Secur. **28**(1–2), 18–28 (2009)
91. S. Garg, K. Kaur, N. Kumar, G. Kaddoum, A.Y. Zomaya, R. Ranjan, A hybrid deep learning-based model for anomaly detection in cloud datacenter networks. IEEE Trans. Netw. Serv. Manage. **16**(3), 924–935 (2019)
92. K. Giotis, C. Argyropoulos, G. Androulidakis, D. Kalogeras, V. Maglaris, Combining openflow and sflow for an effective and scalable anomaly detection and mitigation mechanism on sdn environments. Comput. Netw. **62**, 122–136 (2014)
93. Y.H. Goo, K.S. Shim, S.K. Lee, M.S. Kim, Payload signature structure for accurate application traffic classification, in *2016 18th Asia-Pacific Network Operations and Management Symposium (APNOMS)* (IEEE, Piscataway, 2016), pp. 1–4
94. Google/cluster-data (2009). https://github.com/google/cluster-data
95. Google transparency report (2021). https://transparencyreport.google.com/
96. A. Grishina, M. Chinnici, A.L. Kor, E. Rondeau, J.P. Georges, A machine learning solution for data center thermal characteristics analysis. Energies **13**(17), 4378 (2020)
97. J. Guo, Z . Zhu, When deep learning meets inter-datacenter optical network management: advantages and vulnerabilities. J. Lightw. Technol. **36**(20), 4761–4773 (2018)
98. C. Guo, L. Yuan, D. Xiang, Y. Dang, R. Huang, D. Maltz, Z. Liu, V. Wang, B. Pang, H. Chen, et al., Pingmesh: a large-scale system for data center network latency measurement and analysis, in *Proceedings of the 2015 ACM Conference on Special Interest Group on Data Communication* (2015), pp. 139–152
99. Z. Guo, Y. Xu, R. Liu, A. Gushchin, K.y. Chen, A. Walid, H.J. Chao, Balancing flow table occupancy and link utilization in software-defined networks. Future Gen. Comput. Syst. **89**, 213–223 (2018)
100. GWA-T-12 bitbrains (2021). http://gwa.ewi.tudelft.nl/datasets/gwa-t-12-bitbrains
101. GWA-T-13 materna (2021). http://gwa.ewi.tudelft.nl/datasets/gwa-t-13-materna
102. K. Haghshenas, A. Pahlevan, M. Zapater, S. Mohammadi, D. Atienza, Magnetic: Multi-agent machine learning-based approach for energy efficient dynamic consolidation in data centers. IEEE Trans. Serv. Comput. **15**(1), 30–44 (2019)

103. E.L. Hahne, Round-robin scheduling for max-min fairness in data networks. IEEE J. Sel. Areas Commun. **9**(7), 1024–1039 (1991)

104. M. Hall, E. Frank, G. Holmes, B. Pfahringer, P. Reutemann, I.H. Witten, The weka data mining software: an update. ACM SIGKDD Explor. Newslett. **11**(1), 10–18 (2009)

105. Y. Han, J. Li, D. Hoang, J.H. Yoo, J.W.K. Hong, An intent-based network virtualization platform for SDN, in *2016 12th International Conference on Network and Service Management (CNSM)* (IEEE, Piscataway, 2016), pp. 353–358

106. C. Hardegen, B. Pfülb, S. Rieger, A. Gepperth, S. Reißmann, Flow-based throughput prediction using deep learning and real-world network traffic, in *2019 15th International Conference on Network and Service Management (CNSM)* (IEEE, Piscataway, 2019), pp. 1–9

107. C. Hardegen, B. Pfülb, S. Rieger, A. Gepperth, Predicting network flow characteristics using deep learning and real-world network traffic. IEEE Trans. Netw. Serv. Manage. **17**(4), 2662–2676 (2020)

108. H. Hashemi, K. Abdelghany, End-to-end deep learning methodology for real-time traffic network management. Comput.-Aided Civil Infrastruct. Eng. **33**(10), 849–863 (2018)

109. K. He, E. Rozner, K. Agarwal, Y. Gu, W. Felter, J. Carter, A. Akella, AC/DC TCP: virtual congestion control enforcement for datacenter networks, in *Proceedings of the 2016 ACM SIGCOMM Conference* (2016), pp. 244–257

110. A. Headquarters, CISCO data center infrastructure 2.5 design guide. CISCO validated design I (2007)

111. H. Herodotou, B. Ding, S. Balakrishnan, G. Outhred, P. Fitter, Scalable near real-time failure localization of data center networks, in *Proceedings of the 20th ACM SIGKDD International Conference on Knowledge Discovery and Data Mining* (2014), pp. 1689–1698

112. C.Y. Hong, M. Caesar, P.B. Godfrey, Finishing flows quickly with preemptive scheduling. ACM SIGCOMM Comput. Commun. Rev. **42**(4), 127–138 (2012)

113. Y. Hong, X. Hong, J. Chen, Neural network-assisted routing strategy selection for optical datacenter networks, in *Asia Communications and Photonics Conference, Optical Society of America* (2020), pp. S3C–3

114. P. Huang, C. Guo, L. Zhou, J.R. Lorch, Y. Dang, M. Chintalapati, R. Yao, Gray failure: the achilles' heel of cloud-scale systems, in *Proceedings of the 16th Workshop on Hot Topics in Operating Systems* (2017), pp. 150–155

115. S. Ilager, K. Ramamohanarao, R. Buyya, Thermal prediction for efficient energy management of clouds using machine learning. IEEE Trans. Parallel Distrib. Syst. **32**(5), 1044–1056 (2020)

116. Index of /databases/kddcup99 (1999). http://kdd.ics.uci.edu/databases/kddcup99/

117. InitRoot, InitRoot/NSLKDD-dataset (2017). https://github.com/InitRoot/NSLKDD-Dataset

118. Intent-based networking (2017). https://blogs.gartner.com/andrew-lerner/2017/02/07/intent-based-networking/

119. Intent-based networking (IBN) (2017). https://www.cisco.com/c/en/us/solutions/intent-based-networking.html

120. Intent-driven network (2018). https://e.huawei.com/uk/solutions/enterprise-networks/intelligent-ip-networks

121. W. Iqbal, J.L. Berral, D. Carrera, et al., Adaptive sliding windows for improved estimation of data center resource utilization. Future Gen. Comput. Syst. **104**, 212–224 (2020)

122. V. Jeyakumar, M. Alizadeh, Y. Geng, C. Kim, D. Mazières, Millions of little minions: using packets for low latency network programming and visibility. ACM SIGCOMM Comput. Commun. Rev. **44**(4), 3–14 (2014)

123. H. Jiang, Q. Li, Y. Jiang, G. Shen, R. Sinnott, C. Tian, M. Xu, When machine learning meets congestion control: a survey and comparison. Comput. Netw. **192**, 108033 (2021)

124. R. Jin, J. Li, X. Tuo, W. Wang, X. Li, A congestion control method of SDN data center based on reinforcement learning. Int. J. Commun. Syst. **31**(17), e3802 (2018)

125. A. Jobava, A. Yazidi, B.J. Oommen, K. Begnum, On achieving intelligent traffic-aware consolidation of virtual machines in a data center using learning automata. J. Comput. Sci. **24**, 290–312 (2018)

126. U. Johansson, H. Boström, T. Löfström, H. Linusson, Regression conformal prediction with random forests. Mach. Learn. **97**(1–2), 155–176 (2014)
127. L. Jose, L. Yan, M. Alizadeh, G. Varghese, N. McKeown, S. Katti, High speed networks need proactive congestion control, in *Proceedings of the 14th ACM Workshop on Hot Topics in Networks* (2015), pp. 1–7
128. Juniper/open-nti (2016). https://github.com/Juniper/open-nti
129. A. Karthick, E. Ramaraj, R.G. Subramanian, An efficient multi queue job scheduling for cloud computing, in *2014 World Congress on Computing and Communication Technologies* (IEEE, Piscataway, 2014), pp. 164–166
130. R. Kawahara, T. Mori, N. Kamiyama, S. Harada, S. Asano, A study on detecting network anomalies using sampled flow statistics, in *2007 International Symposium on Applications and the Internet Workshops* (IEEE, Piscataway, 2007), p. 81
131. KDD cup 1999 data (1999). http://kdd.ics.uci.edu/databases/kddcup99/kddcup99.html
132. T. Kimura, A. Watanabe, T. Toyono, K. Ishibashi, Proactive failure detection learning generation patterns of large-scale network logs. IEICE Trans. Commun. **102**(2), 306–316 (2019)
133. D. Kliazovich, P. Bouvry, S.U. Khan, Dens: data center energy-efficient network-aware scheduling. Cluster Comput. **16**(1), 65–75 (2013)
134. H. Kuwahara, Y.F. Hsu, K. Matsuda, M. Matsuoka, Dynamic power consumption prediction and optimization of data center by using deep learning and computational fluid dynamics, in *2018 IEEE 7th International Conference on Cloud Networking (CloudNet)* (IEEE, Piscataway, 2018), pp. 1–7
135. H. Kuwahara, Y.F. Hsu, K. Matsuda, M. Matsuoka, Real-time workload allocation optimizer for computing systems by using deep learning, in *2019 IEEE 12th International Conference on Cloud Computing (CLOUD)* (IEEE, Piscataway, 2019), pp. 190–192
136. K. Lee, S. Hong, S.J. Kim, I. Rhee, S. Chong, Slaw: a new mobility model for human walks, in *IEEE INFOCOM 2009* (IEEE, Piscataway, 2009), pp. 855–863
137. Y.L. Lee, D.C. Juan, X.A. Tseng, Y.T. Chen, S.C. Chang, DC-prophet: predicting catastrophic machine failures in datacenters, in *Joint European Conference on Machine Learning and Knowledge Discovery in Databases* (Springer, Berlin, 2017), pp. 64–76
138. K. Lei, M. Qin, B. Bai, G. Zhang, M. Yang, GCN-GAN: a non-linear temporal link prediction model for weighted dynamic networks, in *IEEE INFOCOM 2019-IEEE Conference on Computer Communications* (IEEE, Piscataway, 2019), pp. 388–396
139. D. Lenrow, Intent: don't tell me what to do! (tell me what you want) (2015). https://www.sdxcentral.com/articles/contributed/network-intent-summit-perspective-david-lenrow/2015/02/
140. A. Lerner, J. Skorupa, S. Ganguli, Innovation insight: intent-based networking systems. Tech. Rep., Gartner
141. J. Li, P. Sun, Y. Hu, Traffic modeling and optimization in datacenters with graph neural network. Comput. Netw. **181**, 107528 (2020)
142. Y. Li, H. Liu, W. Yang, D. Hu, X. Wang, W. Xu, Predicting inter-data-center network traffic using elephant flow and sublink information. IEEE Trans. Netw. Serv. Manage. **13**(4), 782–792 (2016)
143. Y. Li, H. Liu, W. Yang, D. Hu, W. Xu, Inter-data-center network traffic prediction with elephant flows, in *NOMS 2016-2016 IEEE/IFIP Network Operations and Management Symposium* (IEEE, Piscataway, 2016), pp. 206–213
144. Y. Li, Y. Wen, D. Tao, K. Guan, Transforming cooling optimization for green data center via deep reinforcement learning. IEEE Trans. Cybern. **50**(5), 2002–2013 (2019)
145. Z. Liao, R. Zhang, S. He, D. Zeng, J. Wang, H.J. Kim, Deep learning-based data storage for low latency in data center networks. IEEE Access **7**, 26411–26417 (2019)
146. Z. Liao, J. Peng, Y. Chen, J. Zhang, J. Wang, A fast Q-learning based data storage optimization for low latency in data center networks. IEEE Access **8**, 90630–90639 (2020)

147. Q. Lin, Z. Gong, Q. Wang, J. Li, Rilnet: a reinforcement learning based load balancing approach for datacenter networks, in *International Conference on Machine Learning for Networking* (Springer, Berlin, 2018), pp. 44–55

148. W.X. Liu, Intelligent routing based on deep reinforcement learning in software-defined datacenter networks, in *2019 IEEE Symposium on Computers and Communications (ISCC)* (IEEE, Piscataway, 2019), pp. 1–6

149. J. Liu, J. Li, G. Shou, Y. Hu, Z. Guo, W. Dai, SDN based load balancing mechanism for elephant flow in data center networks, in 2014 International Symposium on Wireless Personal Multimedia Communications (WPMC) (IEEE, Piscataway, 2014), pp. 486–490

150. B. Liu, Y. Lin, Y. Chen, Quantitative workload analysis and prediction using google cluster traces, in *2016 IEEE Conference on Computer Communications Workshops (INFOCOM WKSHPS)* (IEEE, Piscataway, 2016), pp. 935–940

151. C. Liu, J. Han, Y. Shang, C. Liu, B. Cheng, J. Chen, Predicting of job failure in compute cloud based on online extreme learning machine: a comparative study. IEEE Access **5**, 9359–9368 (2017)

152. N. Liu, Z. Li, J. Xu, Z. Xu, S. Lin, Q. Qiu, J. Tang, Y. Wang, A hierarchical framework of cloud resource allocation and power management using deep reinforcement learning, in *2017 IEEE 37th International Conference on Distributed Computing Systems (ICDCS)* (IEEE, Piscataway, 2017), pp. 372–382

153. W.X. Liu, S.Z. Yu, G. Tan, J. Cai, Information-centric networking with built-in network coding to achieve multisource transmission at network-layer. Comput. Netw. **115**, 110–128 (2017)

154. Z. Liu, D. Gao, Y. Liu, H. Zhang, C.H. Foh, An adaptive approach for elephant flow detection with the rapidly changing traffic in data center network. Int. J. Netw. Manage. **27**(6), e1987 (2017)

155. K. Liu, J. Wang, Z. Liao, B. Yu, J. Pan, Learning-based adaptive data placement for low latency in data center networks, in *2018 IEEE 43rd Conference on Local Computer Networks (LCN)* (IEEE, Piscataway, 2018), pp. 142–149

156. G. Liu, S. Guo, B. Xiao, Y. Yang, SDN-based traffic matrix estimation in data center networks through large size flow identification. IEEE Trans. Cloud Comput. **10**(1), 675–690 (2022)

157. K. Liu, J. Peng, J. Wang, B. Yu, Z. Liao, Z. Huang, J. Pan, A learning-based data placement framework for low latency in data center networks. IEEE Trans. Cloud Comput. **10**, 146–157 (2019)

158. W.X. Liu, J. Cai, Y. Wang, Q.C. Chen, D. Tang, Mix-flow scheduling using deep reinforcement learning for software-defined data-center networks. Internet Technol. Lett. **2**(3), e99 (2019)

159. W.X. Liu, J. Cai, Y. Wang, Q.C. Chen, J.Q. Zeng, Fine-grained flow classification using deep learning for software defined data center networks. J. Netw. Comput. Appl. **168**, 102766 (2020)

160. X. Liu, Y. He, H. Liu, J. Zhang, B. Liu, X. Peng, J. Xu, J. Zhang, A. Zhou, P. Sun, et al., Smart server crash prediction in cloud service data center, in 2020 19th IEEE Intersociety Conference on Thermal and Thermomechanical Phenomena in Electronic Systems (ITherm) (IEEE, Piscataway, 2020), pp. 1350–1355

161. Z. Liu, M. Zhang, X. Zhang, Y. Li, A non-intrusive, traffic-aware prediction framework for power consumption in data center operations. Energies **13**(3), 663 (2020)

162. W.X. Liu, J. Cai, Q.C. Chen, Y. Wang, DRL-R: deep reinforcement learning approach for intelligent routing in software-defined data-center networks. J. Netw. Comput. Appl. **177**, 102865 (2021)

163. J. Ll. Berral, R. Gavaldà, J. Torres, Empowering automatic data-center management with machine learning, in *Proceedings of the 28th Annual ACM Symposium on Applied Computing* (2013), pp. 170–172

164. Y. Lu, X. Fan, L. Qian, Dynamic ECN marking threshold algorithm for TCP congestion control in data center networks. Comput. Commun. **129**, 197–208 (2018)

165. X. Luo, D. Li, Y. Yang, S. Zhang, Spatiotemporal traffic flow prediction with KNN and LSTM. J. Adv. Transp. **2019**, 4145353 (2019)
166. J. Mai, A. Sridharan, C.N. Chuah, H. Zang, T. Ye, Impact of packet sampling on portscan detection. IEEE J. Sel. Areas Commun. **24**(12), 2285–2298 (2006)
167. A. Majidi, X. Gao, S. Zhu, N. Jahanbakhsh, G. Chen, Adaptive routing reconfigurations to minimize flow cost in sdn-based data center networks, in *Proceedings of the 48th International Conference on Parallel Processing* (2019), pp. 1–10
168. A. Majidi, N. Jahanbakhsh, X. Gao, J. Zheng, G. Chen, DC-ECN: a machine-learning based dynamic threshold control scheme for ECN marking in DCN. Comput. Commun. **150**, 334–345 (2020)
169. A. Marahatta, C. Chi, F. Zhang, Z. Liu, Energy-aware fault-tolerant scheduling scheme based on intelligent prediction model for cloud data center, in *2018 Ninth International Green and Sustainable Computing Conference (IGSC)* (IEEE, Piscataway, 2018), pp. 1–8
170. A. Marahatta, Q. Xin, C. Chi, F. Zhang, Z. Liu, PEFS: AI-driven prediction based energy-aware fault-tolerant scheduling scheme for cloud data center. IEEE Trans. Sustain. Comput. **6**, 655–666 (2021)
171. MAWI working group traffic archive (2000). http://mawi.wide.ad.jp/mawi/
172. J. Mbous, T. Jiang, M. Tang, S. Fu, D. Liu, Kalman filtering-based traffic prediction for software defined intra-data center networks. TIIS **13**(6), 2964–2985 (2019)
173. S.A. Mehdi, J. Khalid, S.A. Khayam, Revisiting traffic anomaly detection using software defined networking, in *International Workshop on Recent Advances in Intrusion Detection* (Springer, Berlin, 2011), pp. 161–180
174. A. Merizig, T. Bendahmane, S. Merzoug, O. Kazar, Machine learning approach for energy consumption prediction in datacenters, in *2020 2nd International Conference on Mathematics and Information Technology (ICMIT)* (IEEE, 2020), pp. 142–148
175. A. Mestres, A. Rodriguez-Natal, J. Carner, P. Barlet-Ros, E. Alarcón, M. Solé, V. Muntés-Mulero, D. Meyer, S. Barkai, M.J. Hibbett, et al., Knowledge-defined networking. ACM SIGCOMM Comput. Commun. Rev. **47**(3), 2–10 (2017)
176. R. Mittal, V.T. Lam, N. Dukkipati, E. Blem, H. Wassel, M. Ghobadi, A. Vahdat, Y. Wang, D. Wetherall, D. Zats, Timely: RTT-based congestion control for the datacenter. ACM SIGCOMM Comput. Commun. Rev. **45**(4), 537–550 (2015)
177. A.W. Moore, K. Papagiannaki, Toward the accurate identification of network applications, in *International Workshop on Passive and Active Network Measurement* (Springer, Berlin, 2005), pp. 41–54
178. More google cluster data (2011). http://ai.googleblog.com/2011/11/more-google-cluster-data.html
179. A. Mozo, B. Ordozgoiti, S. Gómez-Canaval, Forecasting short-term data center network traffic load with convolutional neural networks. PloS One **13**(2), e0191939 (2018)
180. A. Murali, N.N. Das, S.S. Sukumaran, K. Chandrasekaran, C. Joseph, J.P. Martin, Machine learning approaches for resource allocation in the cloud: critical reflections, in *2018 International Conference on Advances in Computing, Communications and Informatics (ICACCI)* (IEEE, Piscataway, 2018), pp. 2073–2079
181. J.F. Murray, G.F. Hughes, K. Kreutz-Delgado, D. Schuurmans, Machine learning methods for predicting failures in hard drives: a multiple-instance application. J. Mach. Learn. Res. **6**(5), 783–816 (2005)
182. K. Nagaraj, D. Bharadia, H. Mao, S. Chinchali, M. Alizadeh, S. Katti, Numfabric: fast and flexible bandwidth allocation in datacenters, in *Proceedings of the 2016 ACM SIGCOMM Conference* (2016), pp. 188–201
183. D. Narayanan, A. Donnelly, A. Rowstron, Write off-loading: practical power management for enterprise storage. ACM Trans. Storage **4**(3), 1–23 (2008)
184. L. Nie, D. Jiang, L. Guo, S. Yu, H. Song, Traffic matrix prediction and estimation based on deep learning for data center networks, in *2016 IEEE Globecom Workshops (GC Wkshps)* (IEEE, Piscataway, 2016), pp. 1–6

185. X. Nie, Y. Zhao, Z. Li, G. Chen, K. Sui, J. Zhang, Z. Ye, D. Pei, Dynamic TCP initial windows and congestion control schemes through reinforcement learning. IEEE J. Sel. Areas Commun. **37**(6), 1231–1247 (2019)

186. Online network traffic characterization | ONTIC project | FP7 | CORDIS | European commission (2014). https://cordis.europa.eu/project/id/619633

187. R.P. Padhy, M.R. Patra, S.C. Satapathy, Cloud computing: security issues and research challenges. Int. J. Comput. Sci. Inf. Technol. Secur. **1**(2), 136–146 (2011)

188. S. Panda, Energy efficient routing and lightpath management in software defined networking based inter-DC elastic optical networks. Opt. Fiber Technol. **55**, 102128 (2020)

189. D. Pandit, S. Chattopadhyay, M. Chattopadhyay, N. Chaki, Resource allocation in cloud using simulated annealing, in *2014 Applications and Innovations in Mobile Computing (AIMoC)* (IEEE, Piscataway, 2014), pp. 21–27

190. L. Pang, C. Yang, D. Chen, Y. Song, M. Guizani, A survey on intent-driven networks. IEEE Access **8**, 22862–22873 (2020)

191. F. Paolucci, A. Sgambelluri, M. Dallaglio, F. Cugini, P. Castoldi, Demonstration of GRPC telemetry for soft failure detection in elastic optical networks, in *2017 European Conference on Optical Communication (ECOC)* (IEEE, Piscataway, 2017), pp. 1–3

192. L. Parolini, B. Sinopoli, B.H. Krogh, Reducing data center energy consumption via coordinated cooling and load management, in *Proceedings of the 2008 Conference on Power Aware Computing and Systems, HotPower*, vol. 8 (2008), pp. 14–14

193. U. Paul, J. Liu, S. Troia, O. Falowo, G. Maier, Traffic-profile and machine learning based regional data center design and operation for 5G network. J. Commun. Netw. **21**(6), 569–583 (2019)

194. B. Pfülb, C. Hardegen, A. Gepperth, S. Rieger, A study of deep learning for network traffic data forecasting, in *International Conference on Artificial Neural Networks* (Springer, Berlin, 2019), pp. 497–512

195. A. Phanishayee, E. Krevat, V. Vasudevan, D.G. Andersen, G.R. Ganger, G.A. Gibson, S. Seshan, Measurement and analysis of TCP throughput collapse in cluster-based storage systems, in *6th USENIX Conference on File and Storage Technologies* (FAST), vol. 8 (2008), pp. 1–14

196. E. Pinheiro, W. Weber, L.A. Barroso, Failure trends in a large disk drive population, in *USENIX Conference on File & Storage Technologies, San Jose* (2007)

197. L. Portnoy, Intrusion detection with unlabeled data using clustering. Ph.D. Thesis, Columbia University (2000)

198. R. Potharaju, N. Jain, When the network crumbles: an empirical study of cloud network failures and their impact on services, in *Proceedings of the 4th Annual Symposium on Cloud Computing* (2013), pp 1–17

199. D. Powers, Evaluation: from precision, recall and F-measure to ROC, informedness, markedness and correlation. J. Mach. Learn. Tech. **2**(1), 37–63 (2011)

200. J.J. Prevost, K. Nagothu, B. Kelley, M. Jamshidi, Prediction of cloud data center networks loads using stochastic and neural models, in *2011 6th International Conference on System of Systems Engineering* (IEEE, Piscataway, 2011), pp. 276–281

201. R. Prieto, Cisco visual networking index predicts near-tripling of IP traffic by 2020. Cisco (2016)

202. Q. Pu, G. Ananthanarayanan, P. Bodik, S. Kandula, A. Akella, P. Bahl, I. Stoica, Low latency GEO-distributed data analytics. ACM SIGCOMM Comput. Commun. Rev. **45**(4), 421–434 (2015)

203. Y. Qiao, X. Qiu, L. Meng, R. Gu, Efficient loss inference algorithm using unicast end-to-end measurements. J. Netw. Syst. Manage. **21**(2), 169–193 (2013)

204. M. Raghu, B. Poole, J. Kleinberg, S. Ganguli, J. Sohl-Dickstein, On the expressive power of deep neural networks, in *International Conference on Machine Learning*. Proceedings of Machine Learning Research (2017), pp. 2847–2854

205. K. Ramachandran, I. Sheriff, E. Belding, K. Almeroth, Routing stability in static wireless mesh networks, in *International Conference on Passive and Active Network Measurement* (Springer, Berlin, 2007), pp. 73–82
206. Y. Ran, H. Hu, X. Zhou, Y. Wen, Deepee: joint optimization of job scheduling and cooling control for data center energy efficiency using deep reinforcement learning, in *2019 IEEE 39th International Conference on Distributed Computing Systems (ICDCS)* (IEEE, Piscataway, 2019), pp. 645–655
207. H. Rastegarfar, M. Glick, N. Viljoen, M. Yang, J. Wissinger, L. LaComb, N. Peyghambarian, TCP flow classification and bandwidth aggregation in optically interconnected data center networks. J. Opt. Commun. Netw. **8**(10), 777–786 (2016)
208. A. Rayan, Y. Nah, Resource prediction for big data processing in a cloud data center: a machine learning approach. IEIE Trans. Smart Process. Comput. **7**(6), 478–488 (2018)
209. C. Reiss, J. Wilkes, J.L. Hellerstein, Google cluster-usage traces: format+ schema. Google Inc., White Paper (2011), pp. 1–14
210. C. Reiss, A. Tumanov, G.R. Ganger, R.H. Katz, M.A. Kozuch, Heterogeneity and dynamicity of clouds at scale: google trace analysis, in *Proceedings of the third ACM Symposium on Cloud Computing* (2012), pp. 1–13
211. C. Reiss, A. Tumanov, G.R. Ganger, R.H. Katz, M.A. Kozuch, Towards understanding heterogeneous clouds at scale: google trace analysis. Intel Science and Technology Center for Cloud Computing, Tech. Rep. 84 (2012)
212. A.M. Ruelas, C.E. Rothenberg, A load balancing method based on artificial neural networks for knowledge-defined data center networking, in *Proceedings of the 10th Latin America Networking Conference* (2018), pp. 106–109
213. F. Ruffy, M. Przystupa, I. Beschastnikh, Iroko: a framework to prototype reinforcement learning for data center traffic control (2018). arXiv:181209975
214. M.A.S. Saber, M. Ghorbani, A. Bayati, K.K. Nguyen, M. Cheriet, Online data center traffic classification based on inter-flow correlations. IEEE Access **8**, 60401–60416 (2020)
215. B.K. Saha, D. Tandur, L. Haab, L. Podleski, A survey of online failure prediction methods. ACM Comput. Surv. **42**(3), 1–42 (2010)
216. B.K. Saha, D. Tandur, L. Haab, L. Podleski, Intent-based networks: an industrial perspective, in *Proceedings of the 1st International Workshop on Future Industrial Communication Networks* (2018), pp. 35–40
217. F. Salfner, M. Malek, Using hidden semi-Markov models for effective online failure prediction, in *2007 26th IEEE International Symposium on Reliable Distributed Systems (SRDS 2007)* (IEEE, Piscataway, 2007), pp. 161–174
218. K. Sasakura, T. Aoki, M. Komatsu, T. Watanabe, A temperature-risk and energy-saving evaluation model for supporting energy-saving measures for data center server rooms. Energies **13**(19), 5222 (2020)
219. N. Satheesh, M. Rathnamma, G. Rajeshkumar, P.V. Sagar, P. Dadheech, S. Dogiwal, P. Velayutham, S. Sengan, Flow-based anomaly intrusion detection using machine learning model with software defined networking for openflow network. Microprocess. Microsyst. **79**, 103285 (2020)
220. T. Scherer, J. Xue, F. Yan, R. Birke, L.Y. Chen, E. Smirni, Practise–demonstrating a neural network based framework for robust prediction of data center workload, in *2015 IEEE/ACM 8th International Conference on Utility and Cloud Computing (UCC)* (IEEE, Piscataway, 2015), pp. 402–403
221. B. Schroeder, R. Lagisetty, A. Merchant, Flash reliability in production: the expected and the unexpected, in *14th USENIX Conference on File and Storage Technologies (FAST'16)* (2016), pp. 67–80
222. Q. Schueller, K. Basu, M. Younas, M. Patel, F. Ball, A hierarchical intrusion detection system using support vector machine for SDN network in cloud data center, in *2018 28th International Telecommunication Networks and Applications Conference (ITNAC)* (IEEE, Piscataway, 2018), pp. 1–6

223. S. Sen, O. Spatscheck, D. Wang, Accurate, scalable in-network identification of p2p traffic using application signatures, in *Proceedings of the 13th International Conference on World Wide Web* (2004), pp. 512–521

224. Service name and transport protocol port number registry (2021). https://www.iana.org/assignments/service-names-port-numbers/service-names-port-numbers.xhtml

225. D. Shan, F. Ren, Improving ecn marking scheme with micro-burst traffic in data center networks, in *IEEE INFOCOM 2017-IEEE Conference on Computer Communications* (IEEE, Piscataway, 2017), pp. 1–9

226. D. Shan, F. Ren, P. Cheng, R. Shu, C. Guo, Micro-burst in data centers: observations, analysis, and mitigations, in *2018 IEEE 26th International Conference on Network Protocols (ICNP)* (IEEE, Piscataway, 2018), pp. 88–98

227. V.S. Shekhawat, A. Gautam, A. Thakrar, Datacenter workload classification and characterization: an empirical approach, in *2018 IEEE 13th International Conference on Industrial and Information Systems (ICIIS)* (IEEE, Piscataway, 2018), pp. 1–7

228. S. Shen, V. Van Beek, A. Iosup, Statistical characterization of business-critical workloads hosted in cloud datacenters, in *2015 15th IEEE/ACM International Symposium on Cluster, Cloud and Grid Computing* (IEEE, Piscataway, 2015), pp. 465–474

229. H. Shi, C. Wang, LSTM-based traffic prediction in support of periodically light path reconfiguration in hybrid data center network, in *2018 IEEE 4th International Conference on Computer and Communications (ICCC)* (IEEE, Piscataway, 2018), pp. 1124–1128

230. H. Shi, H. Li, D. Zhang, C. Cheng, W. Wu, Efficient and robust feature extraction and selection for traffic classification. Comput. Netw. **119**, 1–16 (2017)

231. A. Shiravi, H. Shiravi, M. Tavallaee, A.A. Ghorbani, Toward developing a systematic approach to generate benchmark datasets for intrusion detection. Comput. Secur. **31**(3), 357–374 (2012)

232. H. Shoukourian, T. Wilde, A. Auweter, A. Bode, Monitoring power data: a first step towards a unified energy efficiency evaluation toolset for HPC data centers. Environ. Model. Softw. **56**, 13–26 (2014)

233. H. Shoukourian, T. Wilde, D. Labrenz, A. Bode, Using machine learning for data center cooling infrastructure efficiency prediction, in *2017 IEEE International Parallel and Distributed Processing Symposium Workshops (IPDPSW)* (IEEE, Piscataway, 2017), pp. 954–963

234. S.K. Singh, A. Jukan, Machine-learning-based prediction for resource (re) allocation in optical data center networks. IEEE/OSA J. Opt. Commun. Netw. **10**(10), D12–D28 (2018)

235. Software-defined access (2017). https://www.cisco.com/c/en/us/solutions/enterprise-networks/software-defined-access/index.html

236. G. Soni, M. Kalra, A novel approach for load balancing in cloud data center, in 2014 IEEE International Advance Computing Conference (IACC) (IEEE, Piscataway, 2014), pp. 807–812

237. P. Sreekumari, J.I. Jung, M. Lee, An early congestion feedback and rate adjustment schemes for many-to-one communication in cloud-based data center networks. Photon. Netw. Commun. **31**(1), 23–35 (2016)

238. N. Sultana, N. Chilamkurti, W. Peng, R. Alhadad, Survey on SDN based network intrusion detection system using machine learning approaches. Peer Peer Netw. Appl. **12**(2), 493–501 (2019)

239. P. Sun, Z. Guo, S. Liu, J. Lan, Y. Hu, QoS-aware flow control for power-efficient data center networks with deep reinforcement learning, in *ICASSP 2020-2020 IEEE International Conference on Acoustics, Speech and Signal Processing (ICASSP)* (IEEE, Piscataway, 2020), pp. 3552–3556

240. P. Sun, Z. Guo, S. Liu, J. Lan, J. Wang, Y. Hu, Smartfct: improving power-efficiency for data center networks with deep reinforcement learning. Comput. Netw. **179**, 107255 (2020)

241. D. Szostak, K. Walkowiak, Machine learning methods for traffic prediction in dynamic optical networks with service chains, in *2019 21st International Conference on Transparent Optical Networks (ICTON)* (IEEE, Piscataway, 2019), pp. 1–4

242. D. Szostak, K. Walkowiak, Application of machine learning algorithms for traffic forecasting in dynamic optical networks with service function chains. Found. Comput. Decis. Sci. **45**(3), 217–232 (2020)

243. Q. Tang, T. Mukherjee, S.K. Gupta, P. Cayton, Sensor-based fast thermal evaluation model for energy efficient high-performance datacenters, in *2006 Fourth International Conference on Intelligent Sensing and Information Processing* (IEEE, Piscataway, 2006), pp. 203–208

244. Y. Tang, H. Guo, T. Yuan, X. Gao, X. Hong, Y. Li, J. Qiu, Y. Zuo, J. Wu, Flow splitter: a deep reinforcement learning-based flow scheduler for hybrid optical-electrical data center network. IEEE Access **7**, 129955–129965 (2019)

245. M. Tavallaee, E. Bagheri, W. Lu, A.A. Ghorbani, A detailed analysis of the kdd cup 99 data set, in 2009 IEEE Symposium on Computational Intelligence for Security and Defense Applications (IEEE, Piscataway, 2009), pp. 1–6

246. S. Telenyk, E. Zharikov, O. Rolik, Modeling of the data center resource management using reinforcement learning, in *2018 International Scientific-Practical Conference Problems of Infocommunications. Science and Technology (PIC S&T)* (IEEE, Piscataway, 2018), pp. 289–296

247. G. Tesauro, N.K. Jong, R. Das, M.N. Bennani, A hybrid reinforcement learning approach to autonomic resource allocation, in *2006 IEEE International Conference on Autonomic Computing* (IEEE, Piscataway, 2006), pp. 65–73

248. Theophilus A. Benson (2016). http://cs.brown.edu/~tab/

249. The self-driving network: sustainable infrastructure (2017). https://www.juniper.net/uk/en/dm/the-self-driving-network/

250. B. Thiruvenkatam, M.B. Mukeshkrishnan, Optimizing data center network throughput by solving TCP incast problem using k-means algorithm. Int. J. Commun. Syst. e4535 (2020). https://doi.org/10.1002/dac.4535

251. K. Thonglek, K. Ichikawa, K. Takahashi, H. Iida, C. Nakasan, Improving resource utilization in data centers using an LSTM-based prediction model, in *2019 IEEE International Conference on Cluster Computing (CLUSTER)* (IEEE, Piscataway, 2019), pp. 1–8

252. M. Tokic, G. Palm, Value-difference based exploration: adaptive control between epsilon-greedy and softmax, in *Annual Conference on Artificial Intelligence* (Springer, Berlin, 2011), pp. 335–346

253. V. Tosounidis, G. Pavlidis, I. Sakellariou, Deep Q-learning for load balancing traffic in SDN networks, in *11th Hellenic Conference on Artificial Intelligence* (2020), pp. 135–143

254. M.M. Toulouse, G. Doljac, V.P. Carey, C. Bash, Exploration of a potential-flow-based compact model of air-flow transport in data centers, in *ASME International Mechanical Engineering Congress and Exposition*, vol. 43864 (2009), pp. 41–50

255. C. Trois, L.C. Bona, L.S. Oliveira, M. Martinello, D. Harewood-Gill, M.D. Del Fabro, R. Nejabati, D. Simeonidou, J.C. Lima, B. Stein, Exploring textures in traffic matrices to classify data center communications, in *2018 IEEE 32nd International Conference on Advanced Information Networking and Applications (AINA)* (IEEE, Piscataway, 2018), pp. 1123–1130

256. C.F. Tsai, Y.F. Hsu, C.Y. Lin, W.Y. Lin, Intrusion detection by machine learning: a review. Exp. Syst. Appl. **36**(10), 11994–12000 (2009)

257. R.M.A. Ujjan, Z. Pervez, K. Dahal, A.K. Bashir, R. Mumtaz, J. González, Towards sFlow and adaptive polling sampling for deep learning based DDoS detection in SDN. Future Gen. Comput. Syst. **111**, 763–779 (2020)

258. UNIBS: data sharing (2009). http://netweb.ing.unibs.it/~ntw/tools/traces/index.php

259. G. Urdaneta, G. Pierre, M. Van Steen, Wikipedia workload analysis for decentralized hosting. Comput. Netw. **53**(11), 1830–1845 (2009)

260. N. Uv, K.K.G. Pillai, Energy management of cloud data center using neural networks, in *2018 IEEE International Conference on Cloud Computing in Emerging Markets (CCEM)* (IEEE, Piscataway, 2018), pp. 85–89

261. F. Uyeda, L. Foschini, F. Baker, S. Suri, G. Varghese, Efficiently measuring bandwidth at all time scales, in *NSDI'11: Proceedings of the 8th USENIX Conference on Networked Systems Design and Implementation* (2011)

262. B. Vamanan, J. Hasan, T. Vijaykumar, Deadline-aware datacenter TCP (D2TCP). ACM SIGCOMM Comput. Commun. Rev. **42**(4), 115–126 (2012)

263. V.K. Veerabathiran, D. Mani, S. Kuppusamy, B. Subramaniam, P. Velayutham, S. Sengan, S. Krishnamoorthy, Improving secured ID-based authentication for cloud computing through novel hybrid fuzzy-based homomorphic proxy re-encryption. Soft Comput. **24**(24), 18893–18908 (2020)

264. N. Viljoen, H. Rastegarfar, M. Yang, J. Wissinger, M. Glick, Machine learning based adaptive flow classification for optically interconnected data centers, in *2016 18th International Conference on Transparent Optical Networks (ICTON)* (IEEE, Piscataway, 2016), pp. 1–4

265. T. Wang, M. Hamdi, Presto: towards efficient online virtual network embedding in virtualized cloud data centers. Comput. Netw. **106**, 196–208 (2016)

266. Y.C. Wang, S.Y. You, An efficient route management framework for load balance and overhead reduction in SDN-based data center networks. IEEE Trans. Netw. Serv. Manag. **15**(4), 1422–1434 (2018)

267. W. Wang, Y. Sun, K. Zheng, M.A. Kaafar, D. Li, Z. Li, Freeway: adaptively isolating the elephant and mice flows on different transmission paths, in *2014 IEEE 22nd International Conference on Network Protocols* (IEEE, Piscataway, 2014), pp. 362–367

268. T. Wang, B. Qin, M. Hamdi, An efficient framework for online virtual network embedding in virtualized cloud data centers, in *2015 IEEE 4th International Conference on Cloud Networking (CloudNet)* (IEEE, Piscataway, 2015), pp. 159–164

269. B. Wang, J. Su, L. Chen, J. Deng, L. Zheng, Effieye: application-aware large flow detection in data center, in *2017 17th IEEE/ACM International Symposium on Cluster, Cloud and Grid Computing (CCGRID)* (IEEE, Piscataway, 2017), pp. 794–796

270. B. Wang, J. Zhang, Z. Zhang, L. Pan, Y. Xiang, D. Xia, Noise-resistant statistical traffic classification. IEEE Trans. Big Data **5**(4), 454–466 (2017)

271. L. Wang, X. Wang, M. Tornatore, K.J. Kim, S.M. Kim, D.U. Kim, K.E. Han, B. Mukherjee, Scheduling with machine-learning-based flow detection for packet-switched optical data center networks. J. Opt. Commun. Netw. **10**(4), 365–375 (2018)

272. T. Wang, Y. Xia, J. Muppala, M. Hamdi, Achieving energy efficiency in data centers using an artificial intelligence abstraction model. IEEE Trans. Cloud Comput. **6**(3), 612–624 (2018). https://doi.org/10.1109/TCC.2015.2511720

273. R. Wang, C. Wang, X. Gao, H. Guo, J. Wu, Neural network based online flow classifier implemented by fpga in optical DCN, in *2019 24th OptoElectronics and Communications Conference (OECC) and 2019 International Conference on Photonics in Switching and Computing (PSC)* (IEEE, Piscataway, 2019), pp. 1–3

274. S. Wang, J. Zhang, T. Huang, T. Pan, J. Liu, Y. Liu, Improving flow scheduling scheme with mix-traffic in multi-tenant data centers. IEEE Access **8**, 64666–64677 (2020)

275. S. Wang, Y.H. Zhu, S.P. Chen, T.Z. Wu, W.J. Li, X.S. Zhan, H.Y. Ding, W.S. Shi, Y.G. Bao, A case for adaptive resource management in Alibaba datacenter using neural networks. J. Comput. Sci. Technol. **35**(1), 209–220 (2020)

276. Y. Watanabe, H. Otsuka, M. Sonoda, S. Kikuchi, Y. Matsumoto, Online failure prediction in cloud datacenters by real-time message pattern learning, in *4th IEEE International Conference on Cloud Computing Technology and Science Proceedings* (IEEE, Piscataway, 2012), pp. 504–511

277. D. Weerasiri, M.C. Barukh, B. Benatallah, Q.Z. Sheng, R. Ranjan, A taxonomy and survey of cloud resource orchestration techniques. ACM Comput. Surv. **50**(2), 1–41 (2017)

278. C. Wilson, H. Ballani, T. Karagiannis, A. Rowtron, Better never than late: meeting deadlines in datacenter networks. ACM SIGCOMM Comput. Commun. Rev. **41**(4), 50–61 (2011)

279. Y. Xiang, K. Li, W. Zhou, Low-rate DDoS attacks detection and traceback by using new information metrics. IEEE Trans. Inf. Forensics Secur. **6**(2), 426–437 (2011)

280. P. Xiao, W. Qu, H. Qi, Z. Li, Detecting DDoS attacks against data center with correlation analysis. Comput. Commun. **67**, 66–74 (2015)

281. P. Xiao, N. Liu, Y. Li, Y. Lu, X.j. Tang, H.w. Wang, M.X. Li, A traffic classification method with spectral clustering in SDN, in *2016 17th International Conference on Parallel and Distributed Computing, Applications and Technologies (PDCAT)* (IEEE, Piscataway, 2016), pp. 391–394

282. G. Xiao, W. Wenjun, Z. Jiaming, F. Chao, Z. Yanhua, An openflow based dynamic traffic scheduling strategy for load balancing, in 2017 3rd IEEE International Conference on Computer and Communications (ICCC) (IEEE, Piscataway, 2017), pp. 531–535

283. K. Xiao, S. Mao, J.K. Tugnait, TCP-DRINC: smart congestion control based on deep reinforcement learning. IEEE Access **7**, 11892–11904 (2019)

284. Y. Xie, D. Feng, F. Wang, X. Zhang, J. Han, X. Tang, OME: an optimized modeling engine for disk failure prediction in heterogeneous datacenter, in *2018 IEEE 36th International Conference on Computer Design (ICCD)* (IEEE, Piscataway, 2018), pp. 561–564

285. C. Xu, K. Wang, M. Guo, Intelligent resource management in blockchain-based cloud datacenters. IEEE Cloud Comput. **4**(6), 50–59 (2017)

286. Y. Xue, D. Wang, L. Zhang, Traffic classification: issues and challenges. in *2013 International Conference on Computing, Networking and Communications (ICNC)* (IEEE, Piscataway, 2013), pp. 545–549

287. N.J. Yadwadkar, Machine learning for automatic resource management in the datacenter and the cloud. Ph.D. Thesis, UC Berkeley (2018)

288. J. Yan, J. Yuan, A survey of traffic classification in software defined networks, in *2018 1st IEEE International Conference on Hot Information-Centric Networking (HotICN)* (IEEE, Piscataway, 2018), pp. 200–206

289. X. Yang, Y. Wang, H. He, C. Sun, Y. Zhang, Deep reinforcement learning for economic energy scheduling in data center microgrids, in *2019 IEEE Power & Energy Society General Meeting (PESGM)* (IEEE, Piscataway, 2019), pp. 1–5

290. H. Yang, X. Zhao, Q. Yao, A. Yu, J. Zhang, Y. Ji, Accurate fault location using deep neural evolution network in cloud data center interconnection. IEEE Trans. Cloud Comput. **10**, 1402–1412 (2022)

291. W. Yang, Y. Qin, Z. Yang, A reinforcement learning based data storage and traffic management in information-centric data center networks. Mobile Netw. Appl. **27**, 266–275 (2022)

292. Z. Yao, Y. Wang, X. Qiu, DQN-based energy-efficient routing algorithm in software-defined data centers. Int. J. Distrib. Sens. Netw. **16**(6), 1550147720935775 (2020)

293. H. Yi, H. Jung, S. Bae, Deep neural networks for traffic flow prediction, in *2017 IEEE International Conference on Big Data and Smart Computing (BigComp)* (IEEE, Piscataway, 2017), pp. 328–331

294. D. Yi, X. Zhou, Y. Wen, R. Tan, Efficient compute-intensive job allocation in data centers via deep reinforcement learning. IEEE Trans. Parallel Distrib. Syst. **31**(6), 1474–1485 (2020)

295. A. Yu, H. Yang, W. Bai, L. He, H. Xiao, J. Zhang, Leveraging deep learning to achieve efficient resource allocation with traffic evaluation in datacenter optical networks, in *2018 Optical Fiber Communications Conference and Exposition (OFC)* (IEEE, Piscataway, 2018), pp. 1–3

296. C. Yu, J. Lan, Z. Guo, Y. Hu, DROM: optimizing the routing in software-defined networks with deep reinforcement learning. IEEE Access **6**, 64533–64539 (2018)

297. A. Yu, H. Yang, T. Xu, B. Yu, Q. Yao, Y. Li, T. Peng, H. Guo, J. Li, J. Zhang, Long-term traffic scheduling based on stacked bidirectional recurrent neural networks in inter-datacenter optical networks. IEEE Access **7**, 182296–182308 (2019)

298. A. Yu, H. Yang, Q. Yao, Y. Li, H. Guo, T. Peng, H. Li, J. Zhang, Scheduling with flow prediction based on time and frequency 2D classification for hybrid electrical/optical intra-datacenter networks, in *Optical Fiber Communication Conference, Optical Society of America* (2019), pp. Th1H–3

299. A. Yu, H. Yang, Q. Yao, K. Zhan, B. Bao, Z. Sun, J. Zhang, Traffic scheduling based on spiking neural network in hybrid E/O switching intra-datacenter networks, in *ICC 2020-2020 IEEE International Conference on Communications (ICC)* (IEEE, Piscataway, 2020), pp. 1–7

300. J. Yuan, Y. Zheng, X. Xie, G. Sun, Driving with knowledge from the physical world, in *Proceedings of the 17th ACM SIGKDD International Conference on Knowledge Discovery and Data Mining* (2011), pp. 316–324

301. M. Zekri, S. El Kafhali, N. Aboutabit, Y. Saadi, DDoS attack detection using machine learning techniques in cloud computing environments, in *2017 3rd International Conference of Cloud Computing Technologies and Applications (CloudTech)* (IEEE, Piscataway, 2017), pp. 1–7

302. Y. Zeng, H. Gu, W. Wei, Y. Guo, $deep-full-range$: a deep learning based network encrypted traffic classification and intrusion detection framework. IEEE Access **7**, 45182–45190 (2019)

303. J. Zerwas, P. Kalmbach, S. Schmid, A. Blenk, Ismael: using machine learning to predict acceptance of virtual clusters in data centers. IEEE Trans. Netw. Serv. Manage. **16**(3), 950–964 (2019)

304. Y. Zhang, M. Roughan, N. Duffield, A. Greenberg, Fast accurate computation of large-scale IP traffic matrices from link loads. ACM SIGMETRICS Perform. Eval. Rev. **31**(1), 206–217 (2003)

305. Y. Zhang, M. Roughan, W. Willinger, L. Qiu, Spatio-temporal compressive sensing and internet traffic matrices, in *Proceedings of the ACM SIGCOMM 2009 Conference on Data Communication* (2009), pp. 267–278

306. J. Zhang, F. Ren, C. Lin, Modeling and understanding TCP incast in data center networks, in *2011 Proceedings IEEE INFOCOM* (IEEE, Piscataway, 2011), pp. 1377–1385

307. J. Zhang, X. Chen, Y. Xiang, W. Zhou, J. Wu, Robust network traffic classification. IEEE/ACM Trans. Netw. **23**(4), 1257–1270 (2014)

308. S. Zhang, Y. Liu, W. Meng, Z. Luo, J. Bu, S. Yang, P. Liang, D. Pei, J. Xu, Y. Zhang, et al., Prefix: switch failure prediction in datacenter networks. Proc. ACM Measur. Anal. Comput. Syst. **2**(1), 1–29 (2018)

309. L. Zhao, P. Shi, Machine learning assisted aggregation schemes for optical cross-connect in hybrid electrical/optical data center networks. OSA Contin. **3**(9), 2573–2590 (2020)

310. S. Zhao, K. Ye, C.Z. Xu, Traffic classification and application identification based on machine learning in large-scale supercomputing center, in *2019 IEEE 21st International Conference on High Performance Computing and Communications; IEEE 17th International Conference on Smart City; IEEE 5th International Conference on Data Science and Systems (HPCC/SmartCity/DSS)* (IEEE, Piscataway, 2019), pp. 2299–2304

311. Q. Zhou, K. Wang, P. Li, D. Zeng, S. Guo, B. Ye, M. Guo, Fast coflow scheduling via traffic compression and stage pipelining in datacenter networks. IEEE Trans. Comput. **68**(12), 1755–1771 (2019)

312. Z. Zhu, P. Fan, Machine learning based prediction and classification of computational jobs in cloud computing centers, in *2019 15th International Wireless Communications & Mobile Computing Conference (IWCMC)* (IEEE, Piscataway, 2019), pp. 1482–1487

313. B. Zhu, G. Wang, X. Liu, D. Hu, S. Lin, J. Ma, Proactive drive failure prediction for large scale storage systems, in *2013 IEEE 29th Symposium on Mass Storage Systems and Technologies (MSST)* (IEEE, Piscataway, 2013), pp. 1–5

314. K. Zhu, G. Shen, Y. Jiang, J. Lv, Q. Li, M. Xu, Differentiated transmission based on traffic classification with deep learning in datacenter, in *2020 IFIP Networking Conference (Networking)* (IEEE, Piscataway, 2020), pp. 599–603

315. S. Zou, J. Huang, J. Wang, T. He, Flow-aware adaptive pacing to mitigate TCP incast in data center networks. IEEE/ACM Trans. Netw. **29**(1), 134–147 (2020)

Chapter 4
Insights, Challenges and Opportunities

Through systematic research and analysis, we found that ML has been gradually introduced and applied to various fields of data center network, and has made certain achievements. However, the current researches are still in its infancy and need to be further improved in various areas. The survey [1] by the Uptime Institute in 2020 confirms our view, stating that ML will not take over data center operations and maintenance at this time. In order to further figure out the current progress of ML application in DCN, in this book we investigate and summarize the popularity of different ML technologies in different DCN fields from different perspectives, as shown in Fig. 4.1. Moreover, based on the statistics of the existing work, we make a further analysis from the aspects of ML technology selection, focuses of DCN fields, and REBEL-3S assessment, and provide some more in-depth insights.

1. **Technology Selection:** ML has been carried out in a series of work in various research areas of DCN. Deep learning has gained the favor of researchers because of its good comprehensive ability, accounting for over 50% of all solutions. Supervised learning and deep reinforcement learning are ranked second and third, respectively. According to our current research, the lack of universality [7–10] and reproducibility [5] are the important reasons why reinforcement learning only ranks third up to now. As for the experimental verification of these schemes, over 35% of schemes were conducted based on simulated data other than real-world data, which lacks convincing results to prove their effectiveness in real-world environments.
2. **Focuses of DCN Fields:** The application progress of ML in different fields of DCN also varies. Currently, the researchers mainly focus on flow prediction, resource management, flow classification, flow scheduling, and load balancing, but pay less attention to route optimization, and congestion control.
3. **REBEL-3S Assessment:** In order to more accurately assess the current research status of data center network intelligence, we further analyzed the existing research work according to the proposed REBEL-3S assessment criteria.

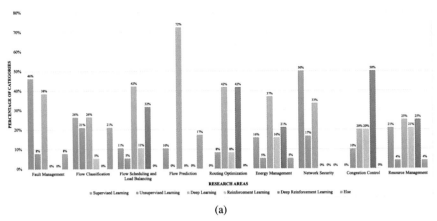

(a)

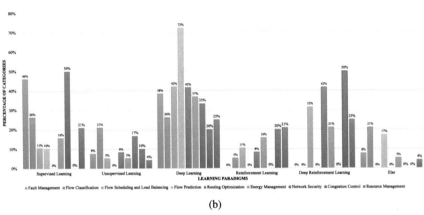

(b)

Fig. 4.1 The current status of intelligence in each research area of DCN. (**a**) From the perspective of research fields of DCN. (**b**) From the perspective of ML algorithms

Besides, we summarize the research progress of ML-based intelligent DCN and draw a vivid Heatmap on Cartesian, as shown in Fig. 4.2, where the "27" in the top-left corner, for example, represents that 27% of the research work in the field of resource management considers RELIABILITY. Clearly, it can be seen that most of the research work has considered bandwidth utilization and stability, where the values of both two columns are above 65%. Comparatively, most of the research work lacks attention to security and reliability, and the results show that most of the values of these two columns remain below 30%. On the other hand, Fig. 4.2 also reflects that the research work in the fields of resource management, flow scheduling and load balancing considers more dimensions of REBEL-3S, and the solutions are more mature.

Overall, most of the existing work still has some unresolved issues, and there are still many opportunities to explore and also a lot of room to improve the

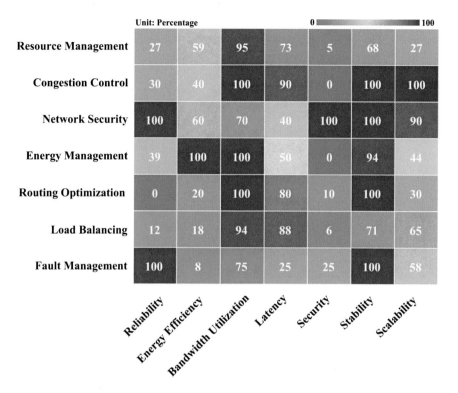

Fig. 4.2 Heatmap on Cartesian of various research fields in accordance with REBEL-3S

level of intelligence. Our view is that the future data center network should be endogenously embedded with intelligence. It is admitted that up to now ML has gained much popularity in various industries, especially in the field of data center network, however, currently its role is more like a tool or module grafted in the system. Whereas, we insist that the intelligence of the future data center network should be an intrinsic natural attribute (Fig. 4.3).

In addition, the network intelligence have been discussed for a long time, but how to define and quantify network intelligence still remains not standardized, and there is not a recognized measurement criterion as well. Huawei has proposed a grading scheme of intelligent networks [2], as shown in Fig. 4.4, which defines six levels of intelligence, ranging from L0 to L5. The L0 intelligence has the ability of auxiliary monitoring, and the execution of all dynamic tasks still depends on manual operations, while the highest L5 intelligence realizes a fully autonomous network with full life cycle closed loop automation capabilities across multiple services and domains. However, this grading scheme still does not provide a formulaic quantitative method to directly and accurately quantify the intelligence level of a network. Therefore, there is still a strong need to further explore a specific

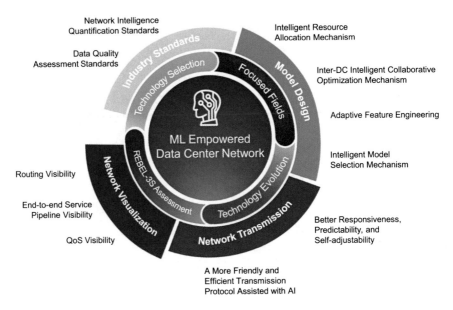

Fig. 4.3 Huawei's five-level autonomous driving network

	Autonomous networks levels					
Level Definition	L0: Manual Operation & Maintenance	L1: Assisted Operation & Maintenance	L2: Partial Autonomous Network	L3: Conditional Autonomous Network	L4: High Autonomous Network	L5: Full Autonomous Network
Execution	P	P/S	S	S	S	S
Awareness	P	P	P/S	S	S	S
Analysis	P	P	P	P/S	S	S
Decision	P	P	P	P/S	S	S
Intent/Experience	P	P	P	P	P/S	S
Applicability	N/A	Select scenarios				All scenarios

P:Personnel S:Systems

Fig. 4.4 Huawei's five-level autonomous driving network (Table modified based on *Huawei ADN Solution White Paper*)

quantitative formula for grading the degree of intelligence similar to the Shannon formula.

Finally, before concluding this book, we will further discuss the challenges and opportunities of data center network intelligence from four aspects: industry standards, model design, network transmission and network visualization, as shown in Fig. 4.3.

4.1   Industry Standards

4.1.1   Network Intelligence Quantification Standards

As aforementioned, because the research on ML-based intelligent data center is still in the initial stage, both the academia and industry have not formulated a specific quantitative standard to assess the network intelligence level. Although some leading high-tech companies (e.g. Huawei) have proposed some directive principles for grading network intelligence levels, which only defines the characteristics and capabilities of the network with different intelligence levels, this is far from being adequate. The ultimate goal is to design a mathematical formula like intelligence quantification method with fairness and accuracy, though there is a long way to go.

4.1.2   Data Quality Assessment Standards

The quality of the source data includes authenticity, validity, diversity, and time-liness. Simulated data lack convincingness, scenario-specific generated data lack universal validity, data containing only a few feature information is challenging to improve the accuracy of predictions, and antiquated historical data lose timeliness having little value. Some existing solutions did not provide any information about the data source, making it difficult to examine the quality of data sets and the validity of experiments. It is necessary to call on researchers to develop a data quality assessment standard as soon as possible, where a quantifiable data quality assessment standard will greatly help enhance the convincingness of experimental results and advance the network intelligence process to some extent.

4.2   Model Design

4.2.1   Intelligent Resource Allocation Mechanism

Data center networks need to intelligently perceive scenarios and services, and reasonably consider the lifecycle of resource management, i.e., resource prediction, allocation, utilization, integration, and recovery, under the security conditions. However, most the existing solutions mainly focus on the resource prediction, allocation and utilization optimization, and there is little research on the resource fragment integration and resource recovery.

4.2.2 Inter-DC Intelligent Collaborative Optimization Mechanism

The inter-DC network optimization is also a very important but more complex research topic, where the optimization usually requires close collaborations among multiple data centers. Thereby, how to achieve efficient collaboration among different intelligent models of different data centers has become a big challenge. Ideally, all separate models can be globally trained based on a complete set of all DC's data, however, normally local data cannot be transferred freely across data centers due to privacy and bandwidth overhead issues. Hence, it will be a good research opportunity to explore efficient methods to achieve an efficient collaboration of inter-DC intelligent models on the premise of ensuring data privacy and security.

4.2.3 Adaptive Feature Engineering

Feature engineering largely affects the ultimate effect of machine learning models. Usually, the feature engineering in ML models is specially designed for a single problem in a specific scenario. However, the richness of data center network layers makes the data collected at each layer vary greatly, and the diversity of services also makes the corresponding feature selection different. How to make feature engineering adaptive to network scenarios and service types under the above complex environment is a key challenge for feature engineering in DCN.

4.2.4 Intelligent Model Selection Mechanism

Without doubt, there is no one universal learning model that works for all scenarios, and every model has certain limitations in different scenarios. The highly dynamic nature of the network environment, the diversity of service requirements, the heterogeneity of network data, and the inconsistency of optimization goals make it extremely difficult and time-consuming to select the most suitable machine learning model. This is also one of the key pain points of the application of artificial intelligence in data center networks.

4.3 Network Transmission

As the data center service scenarios become more and more complex, the network scale becomes larger and larger, and the requirements for user experience and service quality become higher and higher, the traditional communication protocols have already not been qualified to cope with these challenges. Thus, the data center network inevitably requires more efficient and intelligent communication protocols to ensure fast convergence, high bandwidth, low latency and no packet loss for network transmissions. However, the new communication protocols proposed in recent years also fail to meet data center requirements of heterogeneous scenarios [6] and have compatibility issues with legacy protocols [3, 4]. There is no doubt that artificial intelligence can help network protocols achieve better responsiveness, predictability, and self-adjusting ability. However, there is still little research on how to achieve a more friendly and efficient transmission protocol assisted with ML, which is an opportunity for future research.

4.4 Network Visualization

Due to the rapid growth of data center networks, the network size has expanded dramatically and the amount of network data generated has increased greatly. As a result, the burden of network supervision is getting heavier. A proven way to accurately monitor and control from the massive amount of data is network visualization. Network visualization is a comprehensive and concise display of network data by means of graphics, and its ability to reduce the burden of traditional network monitoring. Network visualization can help the O&M personnels accurately perceive the network by explicitly presenting the real-time status of the network to them. Currently, the data center network still has the following three aspects of invisibility, which leads to inefficiency of network O&M and optimization.

- **Routing invisible:** Invisible routing makes the transmission changes cannot be reproduced and the changing process cannot be backtracked. This often leads to a tough situation, that is, the user reported one network fault, but when the O&M personnel starts to locate the fault, the fault disappears again, and there is no historical information to query. As a result, the cause of the fault cannot be diagnosed.
- **End-to-end service pipeline invisible:** This leads to the inability to see the actual forwarding path corresponding to the service pipeline, as well as the performance of the forwarding path. As a result, after the network failure occurs, we can only locate the failure hop by hop, which is time-consuming and laborious.
- **QoS invisible:** The service quality is not visible, resulting in the user experience cannot be perceived. The traditional network management tools usually can only provide the performance data of network, but cannot exhibit the quality of the service contents carried by the network. In other words, the network performance and service quality are separated without any correlations, resulting in low efficiency of fault location.

Overall, the research on applying ML to achieve and intelligent data center networking is still at an early stage. There are always many opportunities to further explore the potential and value of applying ML technologies in various fields of data center networks. It can conclude that the network intelligence will inevitably become the future trend of data center network development. In the foreseeable future, ML-based intelligent networking will become the core research direction of the cloud computing, driving the data center network from SDN-enabled automatic network to ML-driven intelligent network.

References

1. 2020 data center industry survey - uptime institute (2020). https://uptimeinstitute.com/resources/asset/2020-data-center-industry-survey
2. Autonomous driving network (ADN)-Huawei (2020). https://carrier.huawei.com/en/adn
3. B. Cronkite-Ratcliff, A. Bergman, S. Vargaftik, M. Ravi, N. McKeown, I. Abraham, I. Keslassy, Virtualized congestion control, in *Proceedings of the 2016 ACM SIGCOMM Conference* (2016), pp. 230–243
4. B. Cronkite-Ratcliff, A. Bergman, S. Vargaftik, M. Ravi, N. McKeown, I. Abraham, I. Keslassy, AC/DC TCP: virtual congestion control enforcement for datacenter networks, in *Proceedings of the 2016 ACM SIGCOMM Conference* (2016), pp. 244–257
5. P. Henderson, R. Islam, P. Bachman, J. Pineau, D. Precup, D. Meger, Deep reinforcement learning that matters, in *Proceedings of the AAAI Conference on Artificial Intelligence*, vol. 32 (2018)
6. G. Judd, Attaining the promise and avoiding the pitfalls of TCP in the datacenter, in *12th USENIX Symposium on Networked Systems Design and Implementation (NSDI'15)* (2015), pp. 145–157
7. M. Lanctot, V. Zambaldi, A. Gruslys, A. Lazaridou, K. Tuyls, J. Pérolat, D. Silver, T. Graepel, A unified game-theoretic approach to multiagent reinforcement learning (2017). arXiv:171100832
8. M. Raghu, A. Irpan, J. Andreas, B. Kleinberg, Q. Le, J. Kleinberg, Can deep reinforcement learning solve erdos-selfridge-spencer games? in *International Conference on Machine Learning*. Proceedings of Machine Learning Research (2018), pp. 4238–4246
9. S. Whiteson, B. Tanner, M.E. Taylor, P. Stone, Protecting against evaluation overfitting in empirical reinforcement learning, in *2011 IEEE Symposium on Adaptive Dynamic Programming and Reinforcement Learning (ADPRL)* (IEEE, Piscataway, 2011), pp. 120–127
10. C. Zhang, O. Vinyals, R. Munos, S. Bengio, A study on overfitting in deep reinforcement learning (2018). arXiv:180406893

Chapter 5
Conclusion

As the core infrastructure, data center provides a strong platform support for cloud computing, and so on. Nevertheless, the rapid growth of its network scale leads to great challenges in network optimization. Fortunately, artificial intelligence provides a promising way to deal these challenges, and it has been successfully employed in various fields of DCNs. Up to now, there have been numerous literature on ML-assisted intelligent data center networking. However, to the best of our knowledge, there is a lack of systematic investigations into these research works. To this end, in this survey book, we comprehensively review the representative research works with in-depth analysis and discussions from various perspectives including flow prediction, flow classification, load balancing, resource management, energy management, routing optimization, congestion control, fault management, and network security. Notably, to better assess the existing works we creatively propose the REBEL-3S quality assessment scheme. Finally, we thoroughly explore the challenges existed in current research and opportunities for future research from various aspects together with our key insights. To sum up, the research on the application of artificial intelligence in data center networks is still in its infancy, but it has aroused the attention of more and more scholars and researchers, and has achieved preliminary research results in many fields. However, there are still many problems and deficiencies in the current research, which remain to be further studied.

Index

A
Adaptive feature engineering, 106

C
Centralized congestion control, 62
Congestion control, 61

D
Data center level, 48
Data collection scenarios, 11
Data collection techniques, 11
Data quality assessment standards, 105
Deep learning-based flow classification, 23
Distributed congestion control, 63

E
Energy management, 46

F
Fault detection, 70
Fault location, 70
Fault management, 64
Fault prediction, 69
Fault self-healing, 70
Feature Engineering, 11
Flow classification, 17
Flow prediction, 15
Fundamentals, 9

I
Industry standards, 105

Intelligent model selection mechanism, 106
Intelligent resource allocation mechanism, 105
Intent-based network (Cisco), 83
Intent-based network (Gartner), 82
Intent-driven network, 80
Inter-DC intelligent collaborative optimization
    mechanism, 106
Inter-DC routing optimization, 55
Intra-DC routing optimization, 55

K
Knowledge-defined network, 80

L
Learning paradigm, 9
Load balancing, 28

M
Model Design, 105

N
Network intelligence quantification standards,
    105
Network level, 47
Network security, 75
Network transmission, 106
Network visualization, 107
New intelligent networking concepts, 79

Q
QoS-oriented resource management, 37

R
REBEL-3S, 13
Reinforcement learning-based flow
      classification, 24
Resource management, 31
Resource prediction-oriented resource
      management, 38
Resource utilization-oriented resource
      management, 38
Routing optimization, 54

S
Self-driving network, 81
Server level, 47
Spatial-dependent modeling, 16

Supervised learning-based flow classification,
      22

T
Task-oriented resource management, 36
Temporal-dependent modeling, 16

U
Unsupervised learning-based flow
      classification, 23

V
Virtual entities-oriented resource management,
      37

Printed in the United States
by Baker & Taylor Publisher Services